Answers on Blueprint Reading

by Roland Palmquist
revised by Thomas J. Morrisey

Rex Miller, Consulting Editor

THEODORE AUDEL & CO.
a division of
G. K. HALL & CO.
Boston/Massachusetts

FOURTH EDITION
FIRST PRINTING

Manufactured in the United States of America.

Palmquist, Roland E.
    Answers on blueprint reading.

    Includes index.
    1. Blue-prints. I. Morrisey, Thomas J., 1931-
II. Title.
T379.P24 1985   604.2′5   85-17721
ISBN 0-8161-1704-7

Contents

CHAPTER 7

CHAPTER 8

CHAPTER 9

CHAPTER 10

CHAPTER 11

CHAPTER 12

oblique lines — vertical lines — line producing movement — direction of pencil movement — arcs and circles — sketching straight lines — two-dimensional figures — freehand ellipses — semifreehand ellipses — three-dimensional figures — isometric projection — sketching machines

CHAPTER 13

CHAPTER 14

CHAPTER 15

CHAPTER 16

CHAPTER 17

CHAPTER 18

CHAPTER 19

CONTENTS

Advantages of computer aided design and drafting — symbols library — dimensioning — special features — software

APPENDIX

Preface

Blueprints provide designers and drafters with a practical method for communicating their ideas to the engineers and craftsmen. Blueprints are used by building contractors, electricians, plumbers, engineers, and other individuals involved in the construction of buildings. In order to be able to read and interpret a blueprint, the user should be familiar with the fundamentals of drafting.

The first portion of this book offers the basic knowledge needed to interpret blueprints. Note that practice drawings enhance understanding of blueprints.

Blueprints are a form of shorthand that not only reflect but also fix on paper the ideas of designers and drafters. The ability to sketch ideas in a logical manner on paper saves considerable time and effort, as well as avoids possible misinterpretations by persons who will follow through on those plans.

Familiarity is required with the symbols used in blueprints to represent various parts and systems. Each trade has its own symbols, and the craftsmen of each trade should learn to recognize the symbols for all other trades. For example, the electrician should understand the plumber's symbols, the plumber should understand the carpenter's symbols, etc. In this way, each craftsman will know what obstacles he may encounter in his work and will then be better prepared to cope with them. A designer may sometimes wish to use symbols other than those recommended for a particular speciality, in which case he should follow these symbols with a "legend" showing what they mean.

Students, apprentices, and active craftsmen will find this book valuable both as a reference and as a study guide. The many illustrations that supplement the practical material will increase understanding of the subject. Because of the dramatic changes in the uses of computers to generate designs and drawings, a chapter covering computer-aided design and drafting has been added to this edition. Computers have introduced a number of changes in how designers and drafters do their work. Computers have taken the drudgery out of designing and drawing. Instead of laboriously drawing and redrawing drawings, designers and drafters can now create designs and drawings electronically. They can also alter their designs and drawings with a touch of a few buttons. Because computer-aided design and drafting systems have become less expensive and easier to use, even low-tech companies have been trading in their drafting tables for computer terminals.

CHAPTER 1

Blueprints

What is a blueprint?

Answer: A blueprint, in the true sense of the term, is a copy of a drawing in which the lines are white and the background is blue.

What other types of prints are often termed blueprints?

Answer: There are prints with blue lines and white background (diazo), and prints with black lines on a white background. There are also prints with brown lines and a white background, called *sepia* prints; a sepia may also be brown lines on a white background on thin paper, or tracing paper. This is often used to make extra tracings.

What are the most common type of prints used today?

Answer: Blue or black lines on a white background.

What is the most common use of a sepia today?

Answer: Sepias might be called, in photographic terms, an extra negative. They may be used in place of the regular tracing. Ordinarily, the sepia lines are reversed in printing, and any changes that you wish to make are made on the blank side. The sepia lines may be erased if any corrections are necessary. Reversing the sepia will give a clearer reproduction on any prints made from the sepia.

How many prints can be made from a single tracing?

Answer: Any number. The number is limited only by the durability of the tracing material used.

What is an aperture card?

Answer: An aperture card is a film of the print, slightly larger than a 35 mm slide, used to reproduce prints at a reduced size for use in the field; it serves as a substitute for the full-size print.

How is this accomplished?

Answer: The film mounted in a card is placed in a machine, which enlarges the film on the aperture card and reproduces a print smaller than the original.

What is a tracing?

Answer: A tracing is a copy from the original, or the original may be made on the tracing, on various grades of a semitransparent paper, film base, or on a specially prepared linen cloth that is coated so that the ink will readily take to the tracing cloth and may be easily erased.

India ink was previously used exclusively for drawing on the above, but a plastic pencil lead is now used that reproduces just as well.

What is tracing paper?

Answer: Tracing paper is a type of paper, often called vellum, that is used in most drafting work. It can be used in making multiple copies of a drawing at low cost. Good tracing paper should have the following qualities: strength, tooth, stability, translucency, and erasability. *Strength* means that the paper should not tear easily. *Tooth* means that the paper has enough surface roughness so that it can hold the lines drawn upon it. *Stability* means that the paper size will not change with alterations in the amount of moisture in the atmosphere. *Translucency* means that the paper will allow light to pass through it; this specific quality contributes to a good reproduction. *Erasability* means that the paper can withstand repeated erasures without damage to the drawing.

What is drawing film?

Answer: Drawing film is film that is used for extremely accurate and expensive drawings. Film can be purchased in either roll or sheet form. It is available in thin sheets with a toothed surface that will hold lines. Film costs considerably more than vellum or tracing paper. However, it has an excellent drawing surface.

What is tracing cloth?

Answer: Tracing cloth is used when greater durability and permanence are required than can be obtained with drawing paper. It is more expensive than drawing paper. Most types of drawing cloth permit the use of both ink and pencil. Some types of tracing cloth are produced that are resistant to moisture from hands or water accidentally dropped on the surface of the drawing.

How many prints can be made from a single tracing?

Answer: The number is limited only by the durability of the tracing itself.

How are the lines put on a tracing?

Answer: A plastic pencil lead is generally used since it is easily erased and reproduces well. If linen tracing cloth is used, a 5H or 6H pencil is used since it erases easily. After the drawing is completed in pencil, it is traced over with black India ink. India ink is used because of its reproduction qualities, durability, erasability, and water resistance.

How is a print made from the tracing?

Answer: The tracing is laid over the sensitized printing paper and exposed to strong light; it is then developed and fixed.

Explain the process of making a print.

Answer: Originally the tracing and printing paper were placed in a printing frame and exposed to sunlight for the length of time necessary to obtain the proper exposure. It was then washed in water and fixed with a special solution, or sometimes one solution did both, and then it was dried. This type of printing is now obsolete.

Printing is now accomplished by feeding the tracing and printing paper into a blueprint machine, such as shown in figure 1.1. The exposure is made by an arc light or strong lamps, and the printing paper is exposed to ammonia fumes which develop it. With the printing machine, the exposure may be kept at the same level. The printing is done in a matter of seconds and comes out dry and uniform. The exposure time depends upon the machine, the type of paper used, etc. These printing machines have variable speeds to give the correct amount of exposure. A tripping device permits the machine to be loaded, set in operation, and left unattended without the possibility of the print being overexposed through the continued burning of the light.

Fig. 1.1. A modern blueprinting machine.

BLUEPRINT READING

It is not necessary to be a drafter in order to be able to read blueprints, but a person familiar with drafting has a considerable advantage over a mechanic who knows nothing about the subject. Therefore, some of the material in this book may help to familiarize the reader with the technique.

What is the advantage of knowing how to read blueprints?

Answer: The mechanic can study the blueprint and determine what is needed. In this way, he does not have to communicate with the foreman for all of the answers and he will be more productive. Certainly the employer is interested in the productive employee.

What does a blueprint show?

Answer: A blueprint is a working drawing of an object or building. It shows the detail of the work and how it is to be done.

When you find a blueprint with some dimensions omitted, what should you do?

Answer: The mechanic should return to the drafter, or person in charge, and secure the dimensions. An attempt to scale the measurement will probably result in errors. The paper may have shrunk; in addition, the drawing is usually made to a scale smaller than the actual size, and errors will creep in when scaling the drawing.

Is scaling ever of value?

Answer: Yes, scaling has value when quickly checking dimensions that do not have to be accurate. For instance, in estimating an electrical wiring job, scaling is practically always done. The law of averages will allow you to reach a reasonably close figure on wiring requirements.

When plans are drawn to scale and you wish to take measurements of piping, conduits, etc., what is the practical method of measurements?

Answer: By the use of a map or plan measure.

Fig. 1.2. Two types of map and plan measures.

What is a map or plan measure?

Answer: A map or plan measurement is an instrument with a watch-type face and a wheel that is run over the print. As the wheel turns, it registers the distance traveled on the dial, which is generally in feet and inches. Two typical map or plan measuring instruments are shown in figure 1.2.

CHAPTER 2

The Scale

What is one of the first items to consider in preparing a drawing or plan?

Answer: The scale to which the drafter will make the drawing or plan.

What is a scale?

Answer: A scale is the ratio between the actual size of the object and the size that it will be drawn. Sometimes a drawing for a small machine part may have to be drawn twice actual size, that is, on a scale of 2:1, so all of the details of the part can be accurately shown. This type of drawing is necessary for parts requiring a number of precision machining operations.

How is the scale usually expressed on a drawing?

Answer: It might be expressed as full size, half size, quarter size, etc., or it might be expressed as 1 inch = 1 foot; 1 inch = 100 feet.; 1 inch = 1,000 feet or any other proportion that might be necessary to use.

How will you know what the scale is?

Answer: The scale is to be placed on the drawing.

On a full-size drawing, the object and drawing are of the same size. When the drawing is marked half size, the object is twice the size of the drawing. Thus, in figure 2.1, the drawing of an object is shown full size, half size, and quarter size. If the object is a cylindrical piece whose height is represented by H and its diameter by D, then these dimensions will be the same for the full-size drawing. That is, $H = h$; $D = d$. For the half-size drawing, $h' = \frac{1}{2} H$; $d = \frac{1}{2} D$. Similarly, for the quarter-size drawing, $h'' = \frac{1}{4} H$; $d'' = \frac{1}{4} D$.

From this it is seen that when the length of an edge on the drawing is the same as the length of the corresponding edge on the object, the drawing is marked *full size* (sometimes *actual size*). If the length of a line on the drawing is half the length of the corresponding line on the object, the drawing is *half size*.

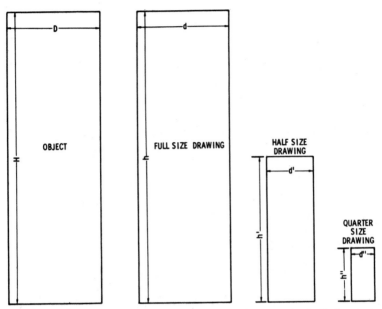

Fig. 2.1. A drawing of an object to different scales: full size, half size, and quarter size.

What are the reasons for using a scale on drawings?

Answer: Usually the scale is smaller than the object, building, etc.; however, this is not always true. It is sometimes advantageous to have a drawing that is larger. In using prints, the size is important since prints are used in the field and, if too large, cannot be easily handled. For example, it would be impossible to have a print as large as a building; thus, it is necessary to reduce the print in size, and this necessitates the use of a scale. In the blueprint of a building the structural, mechanical, and electrical specifications are assembled together. Therefore, by using a scale, the prints will all be equal in size and may be assembled together on one drawing.

Is it necessary that the electrician or mechanic have the complete drawings?

Answer: By having the complete set of prints, the sub-contractor and his men can easily look up any part of the structure and fit their problems into the complete picture, thereby eliminating costly mistakes.

How do architects usually express scale?
Answer: Such as 1 inch = 1 foot; ½ inch = 1 foot; etc.

What does this mean?
Answer: This would indicate that one inch on the drawing would be equal to 1 foot on the actual structure, or ½ inch on the drawing would equal 1 foot on the actual structure respectively.

Do architects and engineers use a different type scale?
Answer: Yes. The architect uses the architects' scale. This is laid out in inches, ⅛'s, ¼'s, ½'s, etc. The engineer uses decimal parts of an inch such as, ¹⁄₁₀ inch, ³⁄₁₀ inch, etc.

In surveying and land plats, which type of scale is used?
Answer: The engineer's scale is used.

In land platting, the figures are usually in feet or miles. How would the scale be expressed?
Answer: Usually it is expressed as 1 inch = 100 feet; 1 inch = 1,000 feet; 1 inch = 1 mile; etc.

Describe a scale used in drafting.
Answer: It is usually 1 foot long. The shape may be triangular or flat. One scale is in inches that are divided into English or metric fractions (see figs. 2.2–3). Regardless of the type of fraction used, the other scales on the ruler will be of the same type.

Describe the architect's scale.
Answer: The architect's scale is flat or triangular. The end inch is divided into twelve equal parts. An architect's scale provides various dimensions for scaling drawings, so that:

1 inch on scale = 1 foot
¾ inch on scale = 1 foot
½ inch on scale = 1 foot

Figure 2.2 shows the details of an architect's flat scale, with ¾ and 1 inch to the foot scales.

How do you lay off a distance of 2′ 6″ on a scale of ¾ inch = 1 foot?

Answer: In figure 2.4, place the ¾ scale with division 2 at the given point A; the zero division on the scale will then be at a distance of 2 feet. Since the end space is divided into twelfths, each division represents one inch on the ¾ scale. Therefore, measuring off six divisions indicates that AB = 2′ 6″. Notice the difference in actual length of this measurement on the 1″ = 1′ scale, shown in figure 2.5.

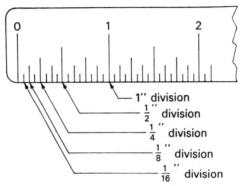

Fig. 2.2. The subdivisions on a full-size scale graduated in the English system of measurement.

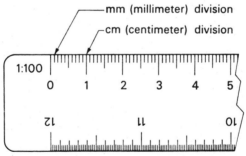

Fig. 2.3. The subdivisions of a full-size scale graduated in the metric system of measurement.

What dimensions are given on a triangular six-scale engineer's scale?

Answer: 10, 20, 30, 40, 50, and 60 divisions to the inch (see fig. 2.6).

Are drawings sometimes made on scales larger than the object?

Answer: Yes. In the case of a very small object, such as a small watch, the drawing may be made two or three times

17

larger for clearness. The scales are indicated as being 2:1, 3:1, etc., meaning twice actual size, three times actual size, etc.

How is the scale affected on drawings reduced in size by photography?

Answer: The scale on the original drawing would not apply to the reproduction.

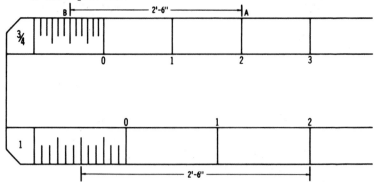

Fig. 2.4. A typical drafter's architect scale; detail showing ¾ and 1 inch to the foot.

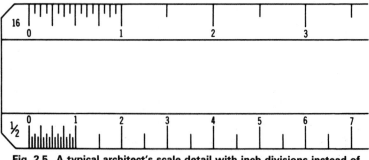

Fig. 2.5. A typical architect's scale detail with inch divisions instead of inches to the foot.

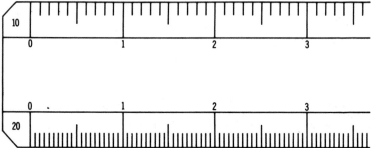

Fig. 2.6. A typical engineer's scale with 10ths and 20ths of inch divisions.

What should be done in such cases?

Answer: The scale on the original drawing (fig. 2.7A) should be crossed out and a graphic scale of proportions corresponding to the reproduction added (fig. 2.7B).

Drafter's scales are generally made of wood or plastic, and are flat or triangular in shape. The architect's scale will always have the scales in inches, halves, quarters, eighths, three quarters, three eighths, etc. There will be one scale in inches, which is usually divided into thirty-seconds of an inch. The other

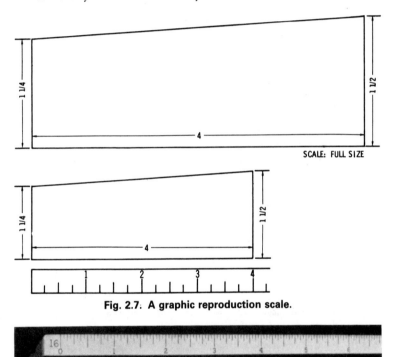

Fig. 2.7. A graphic reproduction scale.

Fig. 2.8. A triangular architect's scale.

Fig. 2.9. A flat architect's scale.

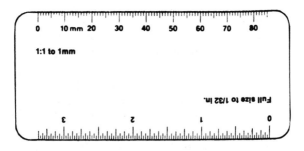

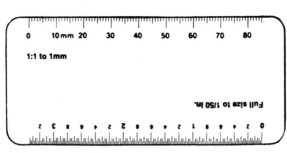

Fig. 2.10. A scale with inches and metric markings.

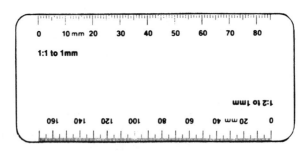

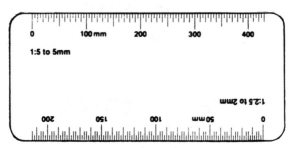

Fig. 2.11. Metric scales with ratio markings.

scales will be to the scale shown by the markings; for example, ½ indicates that one inch is ½ inch on the print.

The engineer's scale uses decimal parts of an inch, and the scales will be indicated as 10, 20, 30, 40, 50, and 60. The 10 scale is divided into inches and tenths of an inch. The 20 scale is one-half size; that is, ½ inch represents 1 inch on the drawing, and again these inches are divided into tenths. The 30 scale means that ⅓ inch is equal to 1 inch on the drawing. Bear in mind that the architect's scale uses the English system, and the engineer's scale the decimal system. Scales come in both flat types and triangular types (see figs. 2.8–9).

Scales are also available that combine inches and metric markings (see fig. 2.10). Metric scales are divided into six ratios. The six ratios are: 1:1 (full scale), 1:2 (half scale), 1:2.5, 1:5, 1:33 ⅓, and 1:75 (see fig. 2.11). These ratios can be reduced or enlarged by dividing and multiplying by 10, which is the common factor of the metric system.

CHAPTER 3

Drafting Instruments

One need not be a drafter in order to read blueprints, but reading is made easier through familiarity with drafting. The intent here is not to give a drafting course, but to provide a knowledge of the instruments used and of some of the technical fundamentals. It is suggested that the reader try drafting or some sketching. Drafting boards and tables come in many sizes and types, but only a few will be illustrated. On practically every job, the mechanic doing the work will have some form of table on which to lay out prints for study and for making needed changes.

For many many years thumbtacks were used to attach the paper to the board, but these made holes and eventually the board had to be discarded. It is now customary to use a piece of tape, similar to masking tape, for this purpose (see fig. 3.1). It leaves no marks and is readily removed when necessary. It also tends to keep the paper tighter. Figure 3.2 illustrates one of the more practical drafting table and board combinations. It is adjustable in height and also in the tilt angle of the top or drawing board. It is much easier to draw on a table top that is tilted. One of the most advanced drafting tables available is the one shown in figure 3.3, manufactured by the Keuffel & Esser Co. This drafting desk has an adjustable drafting board, a desk for sketching or other work that the drafter may wish to do, and drawers for storing instruments.

Contractors and estimators should retain prints of completed jobs for comparison with later assignments. They should also keep at least two sets of prints on a current job. One set is the working drawings for the tradesmen. On the second set, the foreman adds the changes made as the structure was built, to be turned over to the owner to update the original drawings issued. The final drawings will identify any changes that were made during construction. The

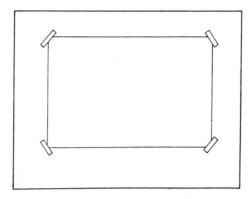

Fig. 3.1. A conventional drawing board with drawing paper taped to the surface.

owner may change the original drawings to reflect the changes, so that his drawings will be fully up to date to assist maintenance men in troubleshooting and designers in making additions to the original installation.

A filing system is required to keep these prints filed. The filing system may be in the form of large shallow drawers (fig. 3.4), tubes (fig. 3.5), or vertical files (fig. 3.6). Any of these types will be satisfactory.

Fig. 3.2. An adjustable drawing stand and board.

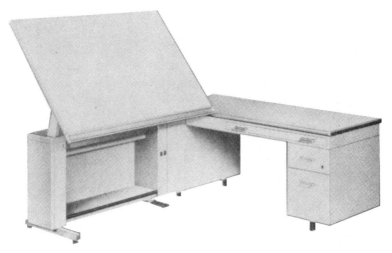

Fig. 3.3. A modern drafting table.

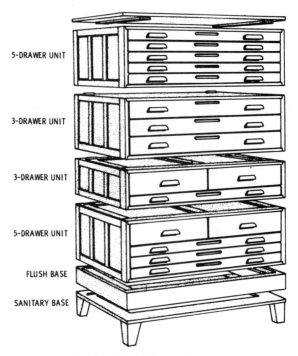

5-DRAWER UNIT

3-DRAWER UNIT

3-DRAWER UNIT

5-DRAWER UNIT

FLUSH BASE

SANITARY BASE

Fig. 3.4. Print filing cabinets

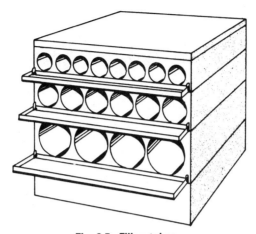

Fig. 3.5. Filing tubes.

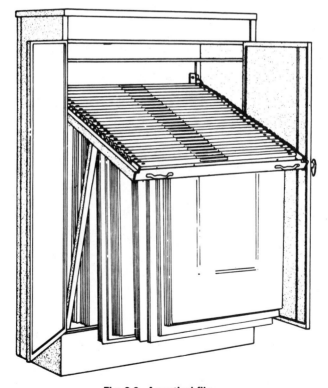

Fig. 3.6. A vertical file.

As to the final "as built" drawings retained by the owner, this and the original drawings are valuable and should be protected from fire and water-sprinkler systems. Fire resistant and waterproof cabinets are provided for mounting the prints vertically.

There are many types of equipment used with drafting tables and boards. The old stand-by is the *T-square* (fig. 3.7). A small version is also available and is handy for field work, where it is used with a clip board for making sketches. The *Jacob's parallel straightedge* is also used by many drafters. It attaches to the drawing board and is adjustable for different angles (fig. 3.8). *Drafting machines*

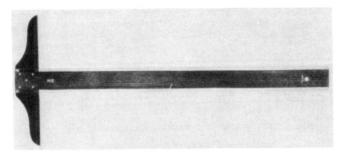

Fig. 3.7. A T-square.

Fig. 3.8. A Jacob's parallel straight edge.

are the most versatile of all. They are adjustable to any angle and have interchangeable scales (figs. 3.9–10).

Probably the next most important drafting instrument is the *triangle*. These are not generally used when a drafting machine is available, but every blueprint reader or drafter should have an assortment of triangles. They come in various sizes and are either 30°, 60°, 90°; 45°, 45°, 90°; or adjustable for different angles (figs. 3.11–13). The lettering holes on the triangles in figures 3.11 and 3.12 give the proper heights for capital and small letters, and the amount of slanting for the letters is also given.

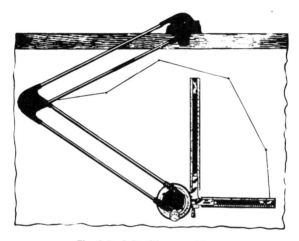

Fig. 3.9. A Drafting machine.

Fig. 3.10. A control head for drafting machine.

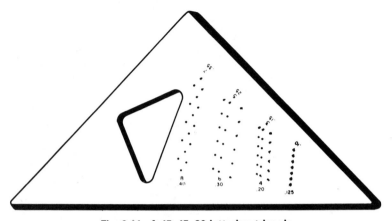

Fig. 3.11. A 45–45–90 lettering triangle.

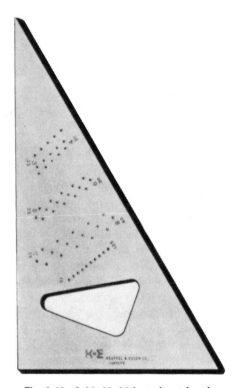

Fig. 3.12. A 30–60–90 lettering triangle.

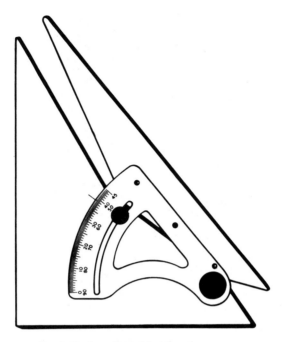

Fig. 3.13. An adjustable triangle.

The *French curve* is another instrument that is valuable to the drafter. It comes in many shapes (figs. 3.14–15) and is used for drawing irregular and curved lines. The drafter and blueprint reader usually have several at their disposal. *Ships* or *graph curves* (fig. 3.16) are often used in place of French curves; they come in various sizes and shapes and they are especially useful in graph work. Graphs are a form of blueprint and have great value in showing trends and figures. More will be covered on graphs later in this book.

Figure 3.17 illustrates a set of drafting instruments for making circles and lines and for use as dividers. The set includes the necessary instruments for pencil and ink work. Figure 3.18 illustrates a proportional divider that is useful for scaling purposes and in making drawings proportionally larger or smaller than the original. Note that the dividers may be set proportionally, which means it may go half-scale, two-times scale, or any other proportion that is desired.

Figure 3.19 illustrates an ink pen for drawing lines. This instrument can be adjusted for different line widths. One side can be moved for cleaning and returned to the correct position without

Fig. 3.14. French curves.

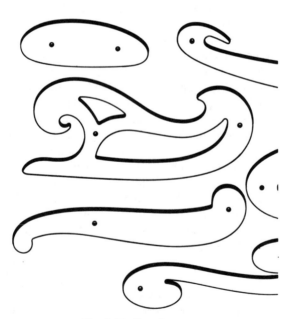

Fig. 3.15. French curves.

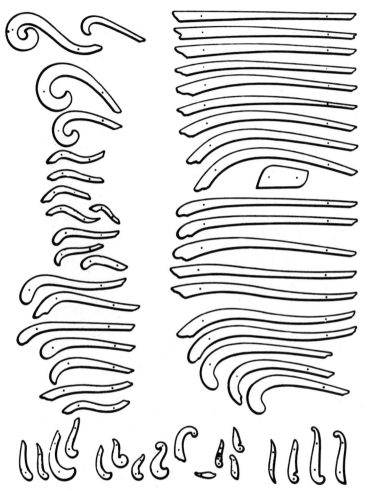

Fig. 3.16. Ships or graph curves.

changing the line width. When considerable drawing is needed, the fountain-type ink pen is valuable (see fig. 3.20). It is usable for both lines and lettering. Self-cleaning is achieved by shaking to clear the point. The pen holds a considerable amount of ink so that it may be used for a great length of time. When the pen is not in use, it is stored with the point up; drainage of ink from the point prevents clogging. The pens come in various sizes for different width lines.

Figure 3.21A illustrates a *Rapidgraph* set that is similar to the pen in figure 3.20 except that there are seven sizes of points in one

Fig. 3.17. Drafting instruments.

Fig. 3.18. Proportional dividers.

Fig. 3.19. Ink ruling pens.

Fig. 3.20. A fountain-type pen for ruling or lettering.

set; the points screw into the handle. The number on the cap indicates the line size, which ranges from 00 (the lightest) to 4 (the heaviest) (see fig. 3.21B). Figures 3.22–24 illustrate a Leroy Lettering set. Figure 3.22 shows the complete set, while figures 3.23–24 show the templates for use with the pen. Figure 3.25 illustrates

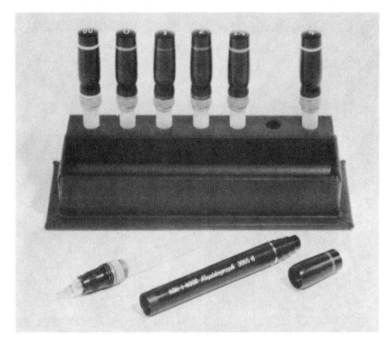

Fig. 3.21A. Rapidograph set.

00 .30	0 .35	1 .50	2 .60	2½ .70	3 .80	3½ 1.00	4 1.20
.012 in.	.014 in.	.020 in.	.024 in.	.028 in.	.031 in.	.039 in.	.047 in.
.30 mm	.35 mm	.50 mm	.60 mm	.70 mm	.80 mm	1.00 mm	1.20 mm

Fig. 3.21B. Point sizes and the line widths produced by each point size.

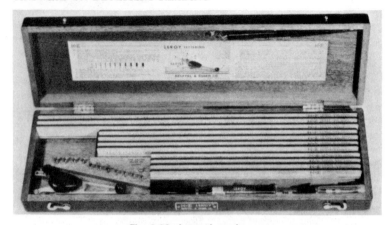

Fig. 3.22. Leroy lettering set.

Fig. 3.23. Lettering templates.

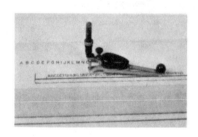

Fig. 3.24. A Leroy lettering pen and template.

ABCDEFGHIJKLMNOPQR
STUVWXYZ·23456789
ABCDEFGHIJKLMNOPQR
STUVWXYZ·123456789

Fig. 3.25. Examples of lettering produced with a lettering template.

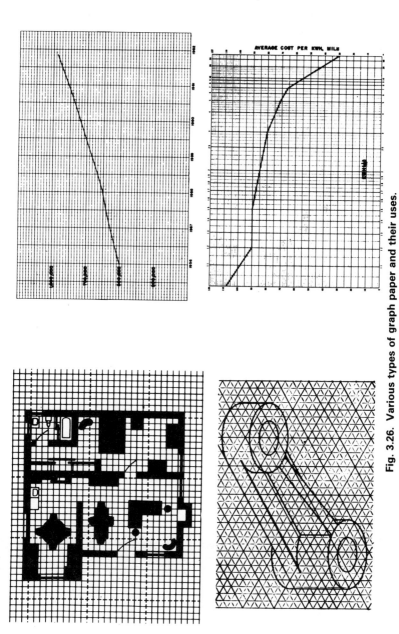

Fig. 3.26. Various types of graph paper and their uses.

the quality of lettering that can be produced with a lettering template. The pen used in the lettering operation has an ink well that Sketching is made simpler in this manner, since the lines can represent any distance that is needed. Thus, the sketch can be accurately completed without having to add the usual dimensions, as shown in figure 3.26.

holds a quantity of ink. The lettering that can be done with this set looks like printing. There are a number of different templates available for many different styles of lettering.

Graphs and sketching are often done on lined paper, commonly known as graph paper, and is available in many styles and patterns.

CHAPTER 4

Conventional Lines

A drawing consists of many different kinds of lines, each having its own purpose. Certain characteristic lines are used to convey different ideas, and the drafting practice has been rather well standardized as to the use of lines to avoid confusion in reading blueprints. A good working drawing is to be made as simple as possible, using only such lines as are necessary to give all of the required information to the mechanic. Moreover, the reader will not have to puzzle over a mass of lines that will complicate the blueprint. The same thing holds true for dimensions and other data. A good drawing is accurate and complete, though simple, and is therefore easily understood ("read") by the mechanic.

The lines generally used on drawings are shown in figure 4.6

What use is made of the heavier lines?
>   *Answer:* Heavier lines are used principally for shading lines.

Where are shading lines placed?
>   *Answer:* Right-hand and lower sides (fig. 4.2).

Name an important use for shading.
>   *Answer:* To more plainly show a cylindrical surface, such as a shaft.

Would the same heavy lines used in this instance be used on a rectangular area as shown in figure 4.2?
>   *Answer:* No. Lighter lines are used that are closer together at the outer surfaces and gradually become farther apart as they approach the center of the shaft. Graduated lines as at *A* in figure 4.3 are preferred to those at *B*.

What important use is made of shading lines?

ANSWERS ON BLUEPRINT READING

Answer: To facilitate the reading of a drawing by bringing out pictorially a cylindrical surface.

For instance, figure 4.3 shows a forging consisting of a rectangular central part having two cylindrical extensions. This is a semipictorial drawing, and the shade lines show to the eye the cylindrical shape of the extensions. A more practical use of these shade lines is shown below in what is called a *sectional view.*

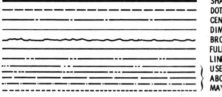

SHADE LINE FOR SHADED DRAWING
DOTTED LINE FOR INVISIBLE SURFACES
CENTER LINE
DIMENSION AND EXTENSION LINES
BROKEN LINE
FULL LINE FOR VISIBLE SURFACES
LINE FOR INDICATING POSITION OF A SECTION
USED FOR CONDITIONS NOT SPECIFIED
ABOVE AND ON GRAPHIC CHARTS, ETC.
MAY BE EITHER LIGHT OR HEAVY.

Fig. 4.1. Various lines used in blueprint drawing.

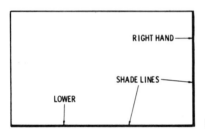

RIGHT HAND

SHADE LINES

LOWER

Fig. 4.2. Illustrating shaded lines.

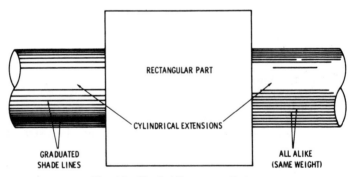

RECTANGULAR PART

CYLINDRICAL EXTENSIONS

GRADUATED
SHADE LINES

ALL ALIKE
(SAME WEIGHT)

Fig. 4.3. Shaded lines on cylinders.

How is a center line illustrated?

Answer: By a long dash and dot line, as shown in figure 4.4.

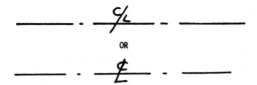

Fig. 4.4. Illustrating a center line.

How does one mark a smoothed or machined surface, such as a cylinder head or bolt?

Answer: The outermost circle and the innermost circle are heavier or outline lines. The fine middle circle indicates the bolt, and the two fine axis lines establish the location of the bolt holes, as shown in figure 4.5.

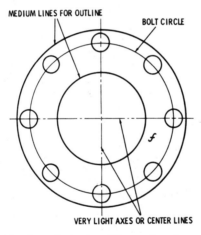

Fig. 4.5. Lower side of a cylinder head illustrating the application of medium and very light lines.

What do cross section lines look like?

Answer: A series of light, solid, parallel lines spaced close together to represent a surface cut by an intersecting plane, as shown in figures 4.6–7.

How does one show a long section of pipe or angle iron without drawing the entire length?

Answer: A broken surface is shown along with the two ends, as illustrated in figure 4.8.

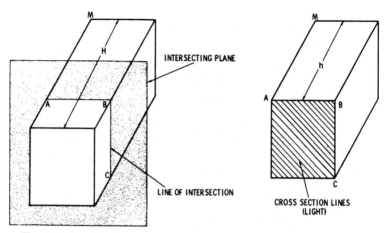

Fig. 4.6. Pictorial view of a block of wood illustrating cross section.

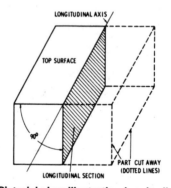

Fig. 4.7. Pictorial view illustrating longitudinal section.

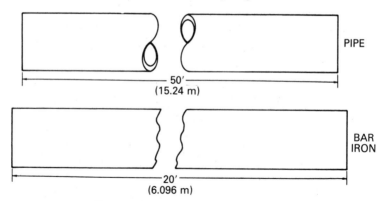

Fig. 4.8. Broken surfaces of pipe and bar.

What is a longitudinal section?

> *Answer:* A longitudinal section is a drawing of an object showing that part cut by a plane passing through its longitudinal axis, as shown in figure 4.7.

> The part cut away is shown in dotted lines. The plane of the section passes through the longitudinal axis and is at 90° to the principal surface of the object, which is the top surface. Strictly speaking, a section shows nothing but the cut surface indicated by the section lines, but for simplicity and to show its position, the pictorial drawing shows part of the object.

What is an oblique section?

> *Answer:* An oblique section is a drawing of an object showing that part cut by a plane which is cutting the object at an *oblique angle*, as shown in figure 4.9.

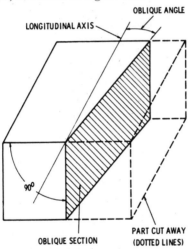

Fig. 4.9. Pictorial view illustrating oblique section.

What is another common name for sectioning?

> *Answer:* Crosshatching.

How are two adjoining, crosshatched pieces identified?

> *Answer:* The crosshatching is in opposite directions, as shown in figure 4.10.

How does one distinguish the junction of two different metals that are to be identified in crosshatching or sectioning?

> *Answer:* One of the metals is represented by alternating solid and broken lines, as in figure 4.11.

41

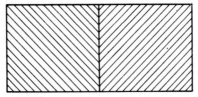

Fig. 4.10. Crosshatching to identify two pieces.

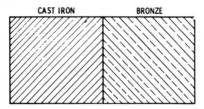

Fig. 4.11. Crosshatching to identify two metals.

DIMENSIONS

Dimensions are items that are extremely important to any drawing. They must be well done, since the final results of all work is based on the dimensions.

What are dimension lines?

Answer: Dimension lines are a means of showing a measurement of length, breadth, height, thickness, or circumference of an object, as shown in figure 4.12.

In figure 2.12A the dimension lines terminate at the dimension limit lines *Aa* and *Bb*. This is the usual method of dimensioning when there is not enough room within the drawing. Figure 2.12B illustrates the method used in avoiding dimension lines by placing the dimension within the outline of the object. When the dimension limit lines are used, the two lines *Aa* and *Bb* (light lines) are drawn, then the dimension line *Dl*, with an arrow at each end touching the limit lines, is added. The dimension line is broken at some point for inserting the dimension figure.

In figure 4.13, *A* represents a dimensioned circle and *B* a dimensioned arc. *D* is the symbol for diameter when dimensioning circles and *R* stands for radius when dimensioning arcs. The small cross in the figure indicates the center of the arc.

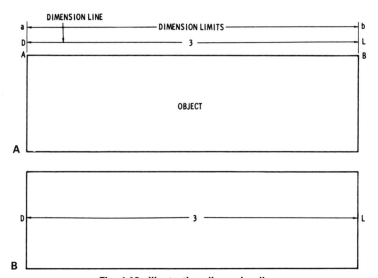

Fig. 4.12. Illustrating dimension lines.

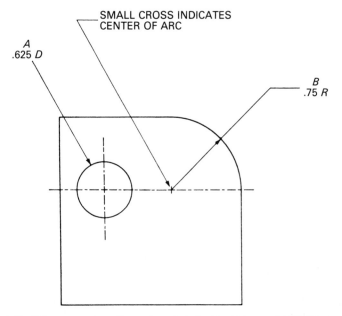

Fig. 4.13. (A) represents a dimensioned circle. The D is used to indicate the diameter of the circle. (B) represents a dimensioned arc. The R is used to indicate the radius of the arc.

43

When does one illustrate a hidden hole or surface?

> *Answer:* This is done by broken lines, as shown in figure 4.14.

What is the correct method of drawing a tangent?

> *Answer:* A tangent is the meeting of a curve or surface and a line at a single point, as shown in figure 4.15A. Figures 4.15B–C show pairs of curves that do not meet at tangents.

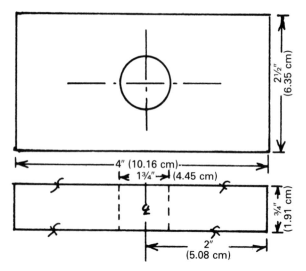

Fig. 4.14. **Illustrating hidden holes.**

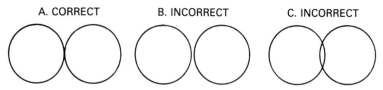

A. CORRECT   B. INCORRECT   C. INCORRECT

Fig. 4.15. **Correct and incorrect methods of making tangents.**

Illustrate the method of dimensioning an angle.

> *Answer:* A method of dimensioning an angle is shown in figure 4.16.

Illustrate the plan and elevation of a cone-shaped object with multisections.

> *Answer:* A cone-shaped object with illustrated multisections is shown in figure 4.17.

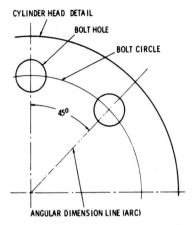

Fig. 4.16. Illustrating the dimensioning of an angle.

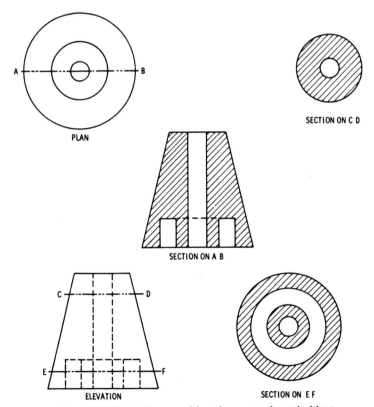

Fig. 4.17. Illustrating a multisection cone-shaped object.

45

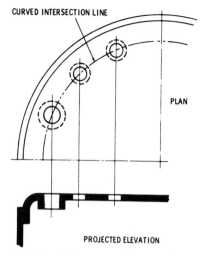

Fig. 4.18. Illustrating curved intersection lines.

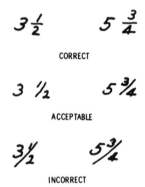

Fig. 4.19. Correct and incorrect way of making fractions.

Illustrate a curved intersection line.

> *Answer:* A curved intersection line is that part of an object cut by an imaginary plane which is one or more sides of an object, as shown in figure 4.18.

Show the correct method of making fractions.

> *Answer:* Figure 4.19 illustrates the correct, acceptable, and incorrect ways of making fractions.

CHAPTER 5

Conventional Representations

.

In addition to the conventional lines presented in the preceding chapters, other standardized devices are used in conveying ideas on blueprints. These consist of various groups of lines, symbols, abbreviations, etc.

REPRESENTATION OF MATERIALS

In the last chapter it was shown how a section is represented by parallel lines drawn close together. By various combinations of light and heavy lines, spacing, etc., cross section lines will not only indicate a section but the kind of metal as well. The various arrangements of section lines to represent these various materials are shown in figure 5.1. Figure 5.2 illustrates standard colors used in blueprint reading.

TOPOGRAPHY

Topography lines are used on maps, subdivision plats, and plats of ground on which a building is to be built, and have other uses as well. On a plat showing a building site, the original topography will be in solid lines and the final topography after leveling will be in dashes. These lines on plans for building sites are essential for determining drainage and landscaping. Topography is generally platted from a starting point of known elevation called a *bench mark.* An overall picture may be obtained that will indicate the proper drainage of surface water from the building site. Figure 5.3 shows a typical illustration of the topography on a parcel of land.

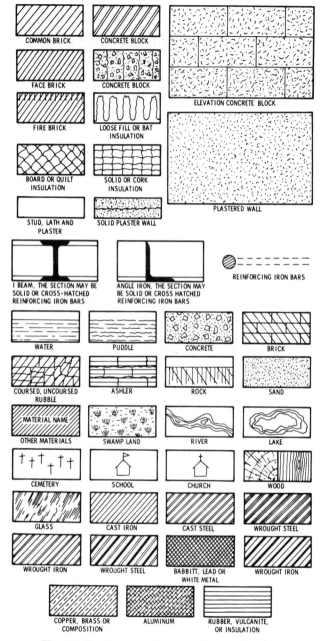

Fig. 5.1. Symbols used for materials and land.

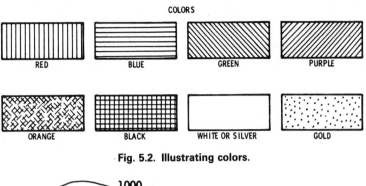

COLORS

Fig. 5.2. Illustrating colors.

Fig. 5.3. Topography lines used on aerial maps.

SCREW THREADS

There are several methods of representing screw threads. The one selected depends on the importance of the drawing and time available. For example, an external thread on a bolt may be shown as in figure 5.4. It takes some time to lay out and draw the threads shown at bolt *A* and they are therefore usually drawn as shown at bolt *B*. The method shown at *C* is used on rush jobs only.

What should be noted about internal threads drawn in sectional views?

Answer: The thread as seen from the back of the section slants opposite to the way it slants on the front or cutaway side (see fig. 5.5). Accordingly, do not think that the threads are

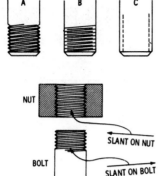

Fig. 5.4. Various methods of representing screw threads.

Fig. 5.5. Nut and bolt threads.

left- or right-hand threads. The internal and external threads are sometimes erroneously made to slant in the same direction.

NUTS

Hexagon nuts and the heads on hexagon bolts are usually shown in drawings with three sides visible. One reason for this is that the clearance space necessary to turn the nut or bolt head should be plainly visible. They are also easier to draw in this position, since the distance AB in figure 5.6 is twice the diameter of the bolt or stud. Sometimes the drafter will show the internal threads by dotted lines. This practice is not necessary as it is understood that there are threads in the inside of the nut.

Figure 5.7 shows the dimensions on a bolt and nut. These dimensions are necessary only when the parts are to be machined. For purchased parts, these dimensions are unnecessary.

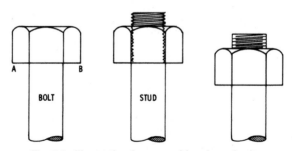

Fig. 5.6. Illustrating hexagonal heads and nuts.

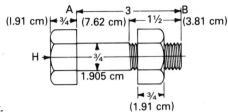

Fig. 5.7. Dimensioning a bolt.

BREAKS FOR CYLINDRICAL PIECES

Long pieces, such as rods, pipes, and pistons, are usually shown only in part to save space on the drawing. Instead of using the break line shown in chapter 4, a more pictorial treatment can be used (see figs. 5.8–9).

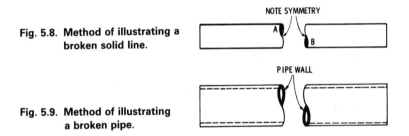

Fig. 5.8. Method of illustrating a broken solid line.

Fig. 5.9. Method of illustrating a broken pipe.

ABBREVIATIONS ON DRAWINGS

A good rule for abbreviations is to use none that would be confusing. If confused as to an abbreviation in reading a blueprint, consider the nature of the piece to be machined and the machining operations, which will be helpful in interpreting the abbreviations. It should be understood that these abbreviations relate only to one part of the subject. For every field, such as carpentry, electrical work, foundry, and shipfitting, there are special conventions. The following are common abbreviations used in blueprint drawings.

Access Door	AD	Aluminum	AL
Access Panel	AP	Anchor	
Acoustic	ACST	Bolt	AB
Aggregate	AGGR	Apartment	APT

Area	A	Cubic Feet	CU FT or FT^3
Area Drain	AD	Detail	DET
Asbestos	ASB	Diagram	DIAG
Asbestos Board	AB	Dining Alcove	DA
Asphalt	ASPH	Dining Room	DR
Asphalt Tile	A Tile	Double Acting	
Automatic Washing		Door	DAD
Machine	AWM	Double Strength	
Basement	BSMT	Glass	DSG
Bathroom	B	Drain	D or DR
Bath Tub	BT	Electric Panel	EP
Beam	BM	End To End	E to E
Bearing Plat	BRG PL	Excavate	EXC
Bedroom	BR	Expansion Joint	EXP JT
Blocking	BLKG	Finished Floor	FIN FL
Blueprint	BP	Firebrick	FRBK
Boiler	BLR	Fireplace	FP
Bolts	BT	Fireproof	FPRF
Book Shelves	BK SH	Flooring	FLG
Boundary	BDY	Flush	FL
Brass	BR	Footing	FTG
Broom Closet	BC	Foundation	FND
Building Line	BL	Frame	FR
Cabinet	CAB	Garage	GAR
Calking	CLKG	Gas	G
Casing	CSG	Gauge	GA
Catch Basin	CB	Gypsum	GYP
Cellar	CEL	Hall	H
Cement Floor	CEM FL	Hardware	HWD
Center	CTR	Hose Bibb	HB
Center to Center	C to C	Hot Air	HA
Center Line	C/L	Hot Water Tank	HWT
Ceramic	CER	I Beam	I
Channel	CHAN	Inside Diameter	ID
Cleanout	CO	Insulation	INS
Clear Glass	CL GL	Iron	I
Closet	CLO	Kitchen	K
Cold Air	CA	Knocked Down	KD
Cold Water	CW	Landing	LDG
Conduit	CND	Lath	LTH
Counter	CTR	Living Room	LR

Main	MN	Section	SECT
Matched and		Sewer	SEW
Dressed	M & D	Shelving	SHELV
Maximum	MAX	Shower	SH
Medicine Cabinet	MC	Single Strength	
Minimum	MIN	Glass	SSG
Miscellaneous	MISC	Sink	S or SK
Mixture	MIX	Soil Pipe	SP
Mortar	MOR	Square Feet	SQ FT or
On Center	OC		FT^2
Pantry	PAN	Stairs	ST
Partition	PARTN	Standard	STD
Plaster	PLAS	Switch	SW or S
Plate	PL	Storage	STG
Porch	P	Telephone	TEL
Precast	PRCST	Thermostat	T or
Prefabricated	PREFAB		THERMO
Pull Switch	PS	Tongue and	
Radiator	RAD	Groove	T&G
Recessed	REC	Unexcavated	UNEXC
Refrigerator	REF	Vent	V
Register	REG	Vinyl Tile	V Tile
Revision	REV	Washroom	WR
Riser	R	Water	W
Rivet	RIV	Water Closet	WC
Room	R or RM	Water Heater	WH
Rubber Tile	R Tile	Weatherstripping	WS
Screen	SCR		

CHAPTER 6

Pictorial Drawings

In approaching the subject of *working drawings*—which are made by the descriptive method of orthographic projection—the reader must first become familiar with the various systems of *pictorial drawing*. The three methods generally used may be classed as *false perspective*, since they show an object approximately as it appears when photographed.

Why are pictorial drawings not drawn in perspective?

Answer: Such drawings present too many difficulties and accordingly take too much time.

Since working drawings are not pictorial drawings, why is it desirable to study the different systems of pictorial drawings?

Answer: On blueprints showing many complicated objects, a pictorial drawing of some detail is included as an aid to reading the blueprint.

PICTORIAL METHODS OF REPRESENTING OBJECTS

The several methods of representing objects in drawings, include

1. Perspective
2. Cabinet projection
3. Modified cabinet projection
4. Isometric projection
5. Anisometric projection

Of these methods, perspective is not covered in this book, as it is not used on blueprints.

What is the advantage of pictorial drawings?

Answer: Most of the object can be seen in one drawing. With the descriptive method used in working drawings only, one side is seen in a single view. Thus, in the pictorial method, the object is seen physically. With the descriptive method, a mental impression of its appearance is conveyed by looking at several views separately.

What are the disadvantages of the pictorial method?

Answer: It requires too much time to make the drawing, it is difficult to fully dimension, some of the dimensions cannot be scaled from the drawings, and all details cannot be fully shown. In cabinet or oblique projection systems, the lines of an object are drawn parallel to three axes.

How are the axes taken in cabinet projection?

Answer: The vertical and horizontal axes lie in a plane intended to appear to the eye as being at right angles to the paper. The axes lie in planes at right angles to each other and are known as the *horizontal, vertical,* and *profile* planes.

How are the axes usually designated?

Answer: As the *X, Y,* and *Z* axis, respectively (fig. 6.1).

At what angle is the Z axis drawn in cabinet projection?

Answer: At 45° (fig. 6.1).

How does one lay out horizontal lines that are parallel to the length of the object ?

Answer: Parallel to the horizontal axis and in their actual sizes.

How about vertical lines parallel to the Y axis?

Answer: They must be drawn parallel to the vertical axis and in their actual sizes.

What should be noted about lines parallel to the Z axis?

Answer: They are laid out one half of their actual sizes.
Problem 1—Draw a cube in cabinet projection, as shown in figure 6.2.

Note that the invisible edges are indicated by dotted lines (as on blueprints). First, draw the three axes, *OX, OY,* and *OZ.* Lay off *OA* and *OC,* on *OX* and *OY,* equal to the side of the given cube, and complete the side by drawing *CB* and *AB.* On *OZ,* lay

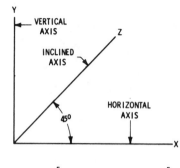

Fig. 6.1. Cabinet projection axes.

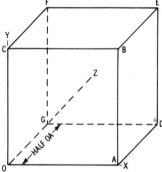

Fig. 6.2. Illustrating a cube.

off *OG* equal to 1/2 *OA*. Through *C*, draw a line parallel to *OZ* and *OG*, and a dotted line parallel to *OY*, giving the lines *CF* and *GF*. Similarly through points *G*, *F*, *A*, and *B* draw parallels to the axes, thus completing the cube.

In the drawing, the face *ABCO* is regarded as lying in the plane of the paper, the face *DEFG* as parallel, and the other faces *ABED* and *OCFG* as perpendicular to the plane of the paper. The edges, which would be invisible if the cube were made of opaque materials such as wood, are represented by dotted lines.

Problem 2—Draw a cylinder in cabinet projection with bases parallel to the plane of the paper. Draw the cylinder with its bases in the *XOY* plan, with the length of the cylinder 3 times the diameter. This is the best position to draw a cylinder, since the bases will be circles and the difficulty of describing ellipse will be avoided. Draw the axes as usual.

With *O* as center (fig. 6.3), and *OA* equal to the radius of the cylinder, scribe a circle. On *OZ*, lay off *OM* equal to 3 times

56

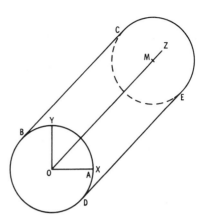

Fig. 6.3. A cylinder in cabinet
projection.

OA (since the length of the cylinder is 3 times the diameter).
With the same radius, scribe a circle through *M* and draw tangents *BC* and *DE*, thus completing the outline of the cylinder.
The portion of the circle *M* between *C* and *F* is shown by dotted
lines because it would be invisible if the cylinder were made of
opaque material.

Problem 3—Draw a prism enclosing a cylinder with its bases parallel to the *YOZ* plane; the length of the cylinder is 3 times the
diameter, as shown in figure 6.4A.

Draw in the cube as directed in problem 1, making $AD = 1/6 \ OA$, as in figure 6.4A. Now, show half of the base *ABED* (fig.
6.4B) in the plane of the paper. Draw diagonals *OB* and *OA* and
scribe the half circle tangent to the sides. Through the intersection of the circle with the diagonal lines, draw diagonal lines, .
draw line *a* and *b*. In figure 6.4A make $Ba = 1/2$ of *Ba* in figure
6.4B and draw lines *a* and *b*, and lines *c* and *d*. Next, draw diago-

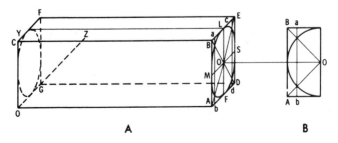

A   B

Fig. 6.4. A prism enclosing a cylinder.

nals *AF* and *BD*, as shown in figure 6.4A. The intersection of lines *a*, *b*, *c*, and *d* with these diagonals will give four points together with points *MLSF*. Construct an ellipse representing the base of the cylinder, as seen in profile, constructing a similar ellipse at the other end, and draw two tangents to the ellipse which completes the outline of the cylinder.

Problem 4—Draw a hexagonal prism scribed in a right cylinder having a given length or altitude (figs. 6.5–6). In the construction of figure 6.5, describe a circle with a diameter equal to the diameter of the cylinder, and inscribe a hexagon. Lay off *OF* and *OC* equal to *Of* and *Oc*. Transfer points *m* and *s*, obtaining *M* and *S*. Draw lines parallel to *OZ* through *M* and *S*, and on these lines lay off $ME = 1/2\ me$; $MA = 1/2\ ma$. Through the points thus obtained, draw in the ellipse *ABCDEF*. Similarly construct the upper ellipse *A'B'C'D'E'F'* at elevation *OO'* from the base and complete the cylinder. Join *AB*, *A'B'*, *BC*, and *B'C'*, etc., and *AA'*, *BB'*, etc., thus completing the outline of the inscribed hexagonal prism.

MODIFIED CABINET PROJECTION

By definition, this system is a method of projection similar to cabinet projection, but differing in that the Z axis is not restricted to 45°. In addition, lines parallel to the Z axis may be drawn on various scales, as half, two thirds, or full size, as shown in figure 6.6.

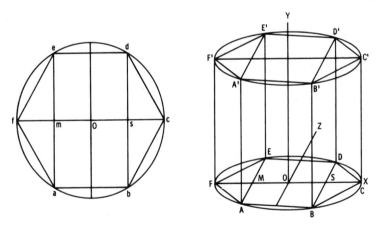

Fig. 6.5. Illustrating a hexagonal prism.

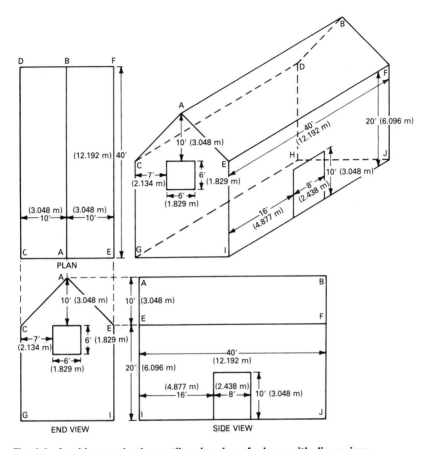

Fig. 6.6. A cabinet projection outline drawing of a barn with dimensions.

What is the object of these modifications?

    Answer: To proportion the drawing to suit the space available.

Name a special provision.

    Answer: Instead of inclining the Z axis to the right, it may be inclined to the left (see figs. 6.7–9).

ISOMETRIC PROJECTION

By definition, the word *isometric* means *equal distances*. Isometric projection is defined as a system of drawing with measurements on

59

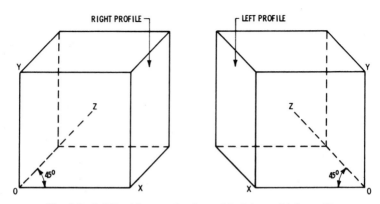

Fig. 6.7. A 45° cabinet projection with right and left profile.

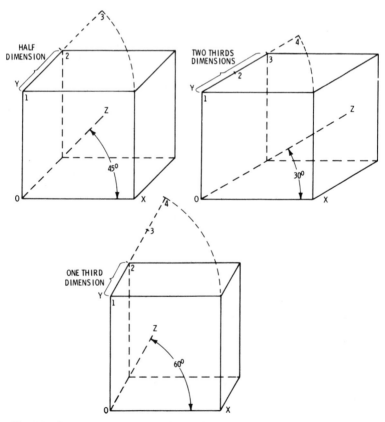

Fig. 6.8. Approved proportions for profile dimensions at 45,° 30,° and 60°.

60

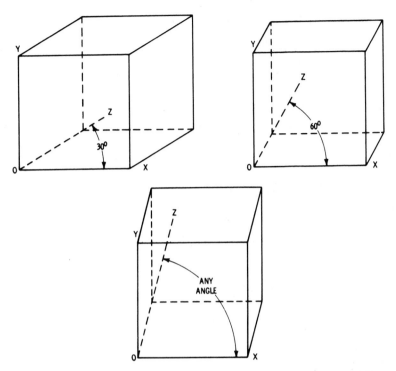

Fig. 6.9. A modified full dimension cabinet projection with axes at 30°, 60°, and at any angle.

an equal scale in every one of three sets of lines 120° apart, and representing the three (XYZ) planes of dimension.

State an important difference between isometric and cabinet drawings.

 Answer: Isometric projection differs from cabinet projection in that none of the three planes lies in the plane of the paper (see fig. 6.10).

How are isometric axes conveniently drawn?

 Answer: With the aid of a T square and a 30° triangle (fig. 6.11A). Figure 6.11B illustrates the step-by-step procedure necessary in the development of an isometric drawing.

How are dimensions laid off on an isometric drawing?

 Answer: They are all laid off on the same scale without any shortening.

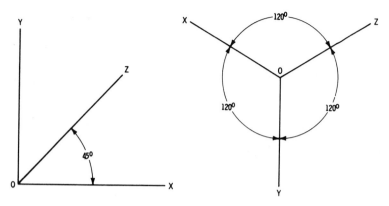

Fig. 6.10. A composition of cabinet and isometric axes.

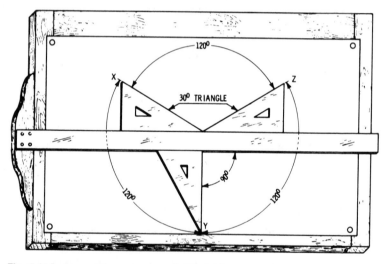

Fig. 6.11A. Isometric axes conveniently laid out to 120° to each other with a T square and a 30° triangle.

What difficulty is encountered in isometric drawing?

Answer: All circles become ellipses. Figure 6.12 illustrates the development of an isometric circle.

Cite another objection to isometric projection.

Answer: Isometric drawings require more space than cabinet drawings (see fig. 6.13).

Problem 5—Draw a prism in isometric projection. First, draw the

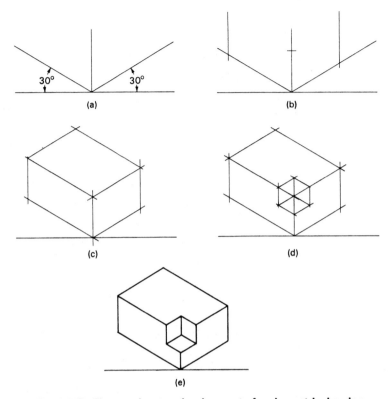

Fig. 6.11B. The step-by-step development of an isometric drawing.

axes OX, OY, and OZ at 120°, as illustrated in figure 6.11A. From point O (fig. 6.14) lay off on the axes just drawn (OA, OB, OC) the length of the side of the cube. Through points ABC thus obtained, draw lines parallel to the axes, giving points DEF, thus completing the visible outline of the cube.

Through DEF draw dotted lines intersecting at G, which gives the invisible outlines of the cube, assuming it to be opaque. An objection to this view is that point G falls behind the line OB, thus the outline of the invisible portion does not appear so well defined as it would in the case of a parallelepipedon (shown as the little figure at the right).

An objection to isometric projection is that, since no projection plane lies in the plane of the paper, it is necessary to construct ellipses to represent circular portions of an object, which requires time and skill.

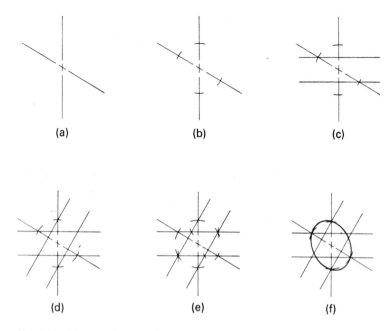

Fig. 6.12. The step-by-step development of an isometric circle. The circle becomes an ellipse.

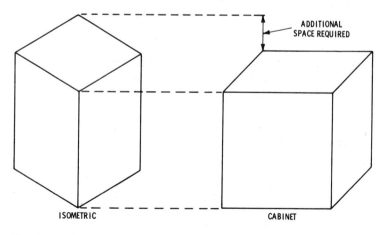

Fig. 6.13. Comparison of isometric and cabinet projection as to relative space required.

64

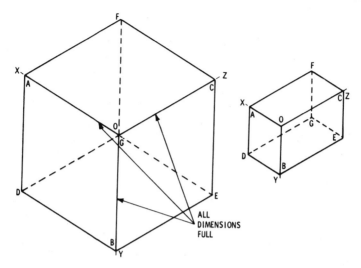

Fig. 6.14. An isometric projection of a cube and of a parallelepipedon.

Problem 6—Draw a horizontal prism with an inscribed cylinder; the length of cylinder will be two times the diameter.

Draw the prism as explained in figure 6.10 and drawn in figure 6.15A making its length twice its side. Then construct the half-end view, as shown in figure 6.15B, and describe the circle, diagonals, and intersecting line *ab*.

The intersections, together with points *MS* and *LE*, of the axial lines through the center give points through which to construct the ellipse. Also construct a similar ellipse at the other end of the prism and join the two ellipses with tangents, thus completing the outline of the inscribed cylinder. Figures 6.16–20 illustrate a number of isometric drawings with proper dimensions.

ANISOMETRIC PROJECTION

This is a modified system. It differs from the cabinet system in that

1. None of its three planes lies in the plane of the paper

2. Its axes lie at different angles

3. Proportionate scales of measurement are used on the different axes

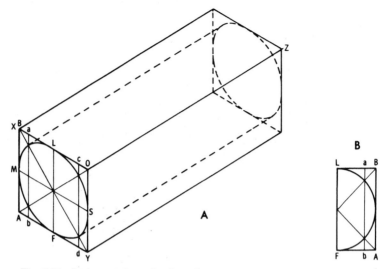

Fig. 6.15. An isometric projection of a horizontal prism with inscribed cylinder.

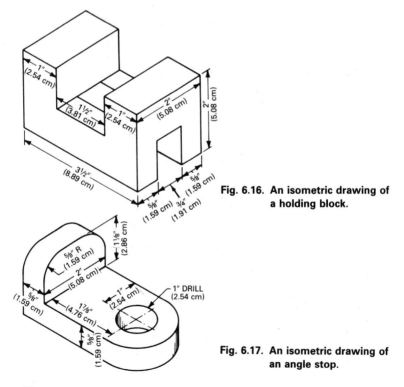

Fig. 6.16. An isometric drawing of a holding block.

Fig. 6.17. An isometric drawing of an angle stop.

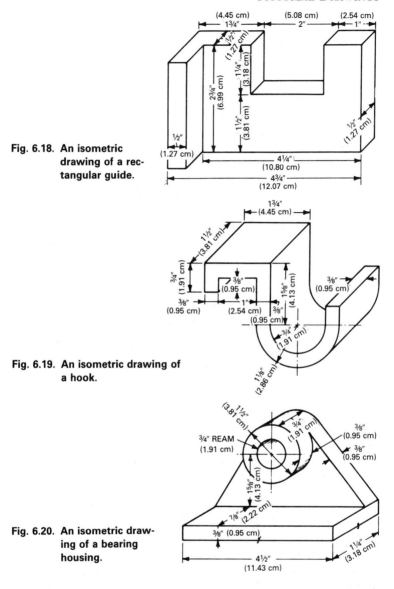

Fig. 6.18. An isometric drawing of a rectangular guide.

Fig. 6.19. An isometric drawing of a hook.

Fig. 6.20. An isometric drawing of a bearing housing.

What is the object of using a different scale and different angles for each axis?

Answer: The angle and proportions for the axes are chosen so that the drawing will be close to true perspective, as shown in figures 6.21A–B. Note that *MS* is to the same scale as *LF*.

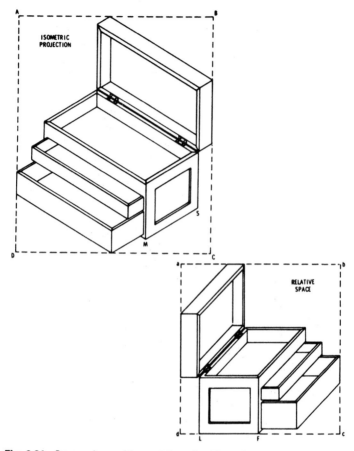

Fig. 6.21. Comparison of isometric and cabinet projection showing relative space required to represent the same object drawn to the same scale.

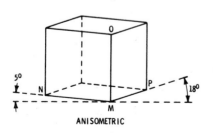

Fig. 6.22. A cube in isometric projection.

Problem 7—Draw a cube in anisometric projection.
In figure 6.22, axis *MN* makes an angle of 5° and axis *MP* an
angle of 18° from the horizontal. From table 6.1, *MO* is drawn to
true scale, *MN* to 7/8 of true scale, and *MP* to 1/2 of true scale.

Table 6.1. Table for Anisometric Projection.

Proportion of Axes			Angles from Horizontal, Degrees	
MO	MN	MP	MN	MP
1	7/8	1/3	5	18
1	3/4	1/2	5	9
1	7/8	3/4	17	25
1	15/16	3/8	4	25
1	15/16	5/16	1 1/2	13

It is seen from this table that the angles and corresponding scales
may be varied to suit the convenience of the drafter, so long as
they do not defeat the purpose of the system, that is, to secure a
closer approach to true perspective.

CHAPTER 7

Geometry of Drafting

There are many phases of drafting, reading blueprints, and the execution of the work that require familiarity with geometry. This chapter is not intended as a course in geometry, but some familiarity with geometrical functions is assumed. Some of the more commonly used geometric figures and related terms are illustrated in figure 7.1.

A *point* is a position in space. When a point is moved, a straight or curved *line* is generated. When a line is moved, a *surface* is generated. Whenever two lines intersect, they form an *angle*. A *circle* is created when a point moves around a second fixed point and maintains a fixed distance from the fixed point. A fixed circle has 360°; one degree is equal to 60′ (minutes) and 1′ is equal to 60″ (seconds). Figure 7.2 illustrates the various terms that are associated with a circle.

The circumference of a circle is equal to a constant of 3.1416 times the diameter, or πd. The area of a circle is πr^2. *Plane surfaces* are areas that contain two intersecting straight lines, such as a triangle (three sides), quadrilateral (four sides), pentagon (five sides), hexagon (six sides), heptagon (seven sides), and octagon (eight sides). *Solids* differ from plane surfaces in that they have three dimensions. To bisect a line AB, as shown in figure 7.3, select a radius greater than half the length of the line AB, and scribe equal arcs using A and B as the center points. Draw a line through the points of intersection of the two arcs, C and D. The point where this line crosses line AB bisects the line AB. To bisect an angle, as shown in figure 7.4, use angle AOB. Use O as a center and draw an arc cutting each side of the angle (points 1 and 2). Using points 1 and 2, draw two arcs cutting each other at C. A line from C to O bisects angle AOB.

To construct a regular polygon (a multisided figure), take a

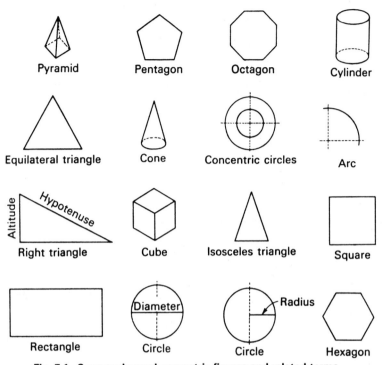

Fig. 7.1. Commonly used geometric figures and related terms.

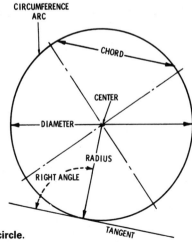

Fig. 7.2. Terms associated with a circle.

71

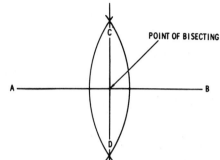

Fig. 7.3. Bisecting a line.

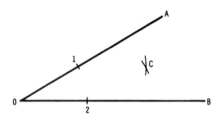

Fig. 7.4. Bisecting an angle.

given side A of the polygon. Draw lines 1 and 2 equal to A, then with 2 as the center, scribe a semicircle with the radius equal to line A. Divide the semicircle into as many parts as there are sides to the polygon, in this instance six. Mark the parts a, b, c, d, e, and f, as shown in figure 7.5. Draw radial lines from point 2 through a, b, c, and d, with d as a center and the radius equal to line A. Draw an arc intersecting line $2c$ at 4, as the center and using the same radius. Cut line $2b$ at 5, with point 1 as center, cut line $2a$ at 6. Connect points 1 to 6, 6 to 5, 5 to 4, 4 to d, and d to 2. This

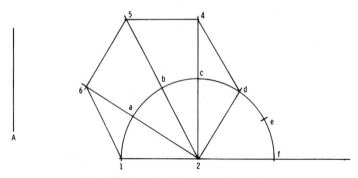

Fig. 7.5. Constructing a regular polygon.

completes the construction of the polygon (in this example a hexagon).

DESCRIPTIVE DRAWING

Descriptive drawing is that branch of drafting concerned with the *descriptive graphic method* of representing points, lines, objects, and the solution of problems relating thereto. It is popularly called *descriptive drawing*. The earnest student will do well to study this chapter as an introduction and preparation for the following chapter on orthographic drawings, which is the method used in making working drawings. Such study will be helpful in understanding orthographic drawings, a knowledge of which is necessary for blueprint reading.

The descriptive method of drawing is based on *parallel projections to a plane by rays perpendicular to the plane*. If the plane is horizontal, the projection is called the *plan* of the figure; if the plane is vertical, the projection is called the *elevation*. The drawings are made to present to the eye the same appearance as the object itself would if it were placed in the proper position.

This descriptive method is also known as *orthographic projection*, and in this chapter some of the basic principles are presented in preparation for the chapters following.

What is projection?

Answer: A projection is a representation of any object on a plane.

How is the line of sight regarded so as to avoid pictorial drawing?

Answer: At an *infinite* distance in a perpendicular drawn to the plane of projection.

Since the point of sight is assumed to be at an infinite distance, what will be the direction of projecting lines drawn from different points of the object being projected?

Answer: They will be parallel with each other and perpendicular to the plane of projection.

What are projection lines?

Answer: Lines of sight.

How do lines of sight proceed from the eye when an object is viewed at a finite distance?

Answer: Radially. The difference between radial and parallel lines of sight (lines of projection) is shown in figure 7.6.

How are projection lines represented on a drawing?
Answer: By light dotted lines.

What mistake is usually made in drawing projection lines?
Answer: Too many lines are drawn.

Why?
Answer: A multiplicity of projection lines is unnecessary and uselessly complicates the drawing, giving it a messy appearance.

PROBLEMS IN PROJECTION

In the first problem two planes are presented at 90° or right angles to each other. The horizontal plane is designated *H*; the vertical plane is represented by *V*.

The *first angle* is above the horizontal and in front of the vertical plane; the *second* is above the horizontal and behind the vertical; the *third* is below the horizontal and behind the vertical; and the *fourth* is below the horizontal and in front of the vertical.

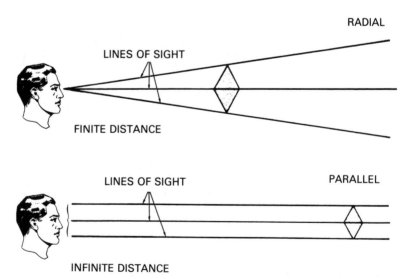

Fig. 7.6. Radial and parallel lines of sight.

It should be noted that the *entire horizontal plane* both in front and behind is *revolved* about the ground line and the revolution is always in a clockwise direction, looking at the two planes from the left side.

Let us consider point *P* (fig. 7.7), which is in the first quadrant. If the horizontal plane is revolved, the projection of point *P* would be revolved downward. When *p* comes into the vertical plane it will be below the vertical projection *p'*. From this it appears that in first-angle projection the plane comes below the *elevation*.

Problem 1—Given the two projections of a point, find the point. In figure 7.7, let *p* and *p'* be the projections of the point in the horizontal and vertical planes, respectively. Draw lines from point *p* to *p'* perpendicular to the *H* and V planes, respectively. The intersection *P* of these lines is at the point required.

Problem 2—Given the projections of the extremities of a line, find the line. This is simply an extension of problem 1.

In figure 7.8 let *ms* and *m's'* be the projections of the extremities of the line in the *H* and V planes, respectively. As in problem 1, use projection to locate the extremities *M* and *S* of the line. Join *M* and *S*, giving *MS*, which is the line required.

Problem 3—Given the traces of a line, find the line.

In figure 7.9, let *ms* and *m's'* (the extremities of the traces) use projection, as in problem 2, to locate the extremities *M* and *S* of the line. Join *M* and *S*, giving *MS*, which is the required line.

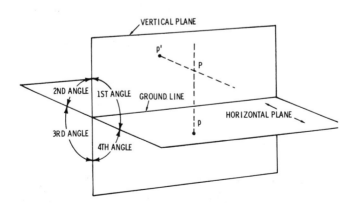

Fig. 7.7. Two projection lines locating a given point.

75

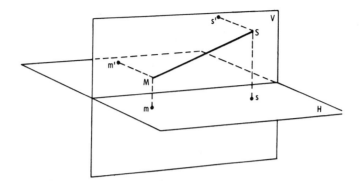

Fig. 7.8. The projection of lines to find a given line.

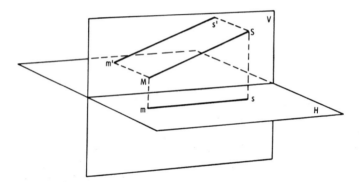

Fig. 7.9. The projection of lines to find a given line that completes an object.

PLANES

A plane is determined by its *two traces,* which are two lines cut on the *projection* planes. If the plane is parallel with the axis, its traces are parallel with the axis. Of these, one may be at infinity; the plane will then cut one of the planes of projection at infinity and will be parallel with it. Thus, a plane parallel with the plane has only one infinite trace, that is, the trace made by its intersection with the V plane. If the plane passes through the axis, both its traces coincide with the axis. This is the only instance in which the representation of the plane by its two traces fails.

Problem 4—Locate the point in which a given straight line extended pierces the planes of projection.

In figure 7.10, let *ms* and *m's'* be the projections of the line. Produce the vertical trace *m's'* until it intersects the ground line at *g'*. At *g'* erect a perpendicular to the ground line in the *H* plane and produce it until it intersects the horizontal trace extended at *h*, the required point of intersection with the *H* plane. By a similar construction, the intersection *h'* with the *V* plane is obtained.

Problem 5—Find the distance between points *M* and *S* in space.

In figure 7.11, let *ms* and *m's'* be the traces of a line joining the two points *MS*. From the ends of the vertical trace *m's'* erect perpendicular lines to the ground line *m'o'* and *s'o'*, cutting the ground line at *o* and *o'* respectively. On the horizontal projection

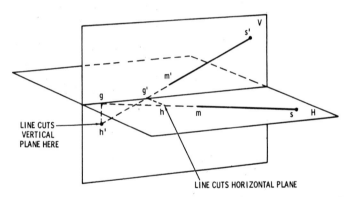

Fig. 7.10. Locating the direction in which a given straight line extends.

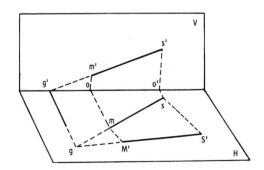

Fig. 7.11. Finding the distance between two points.

ms, erect perpendiculars *mM'* = *om'* and *sS'* = *o's'*. Join the points *M'* and *S'*. The length of this line *M'S'* is equal to the length of the line *MS* (not shown) in space.

In order not to complicate the drawing, the line *MS* in space is not shown. If the construction is accurate, *M'S'* extended will cut *ms* extended at *g* (the point of intersection with the *H* plane), as determined by projecting over from *g'* the intersection of the vertical trace with the axis. The angle diameter is the angle made by *MS* with the *H* plane.

CONIC SECTIONS

By definition, a conic section is *a section cut by a plane passing through a cone*, as shown in figure 7.12. These sections are bounded by well-known curves, and the latter may be any of the following, depending upon the inclination or position of the plane with the axis of the cone:

1. Triangle (plane passes through apex of cone)
2. Circle (plane parallel with base of cone)
3. Ellipse (plane inclined to axis of cone)
4. Parabola (plane parallel with one element of cone)
5. Hyperbola (plane parallel with axis of cone)

These sections will appear as straight lines in elevation, while in a plan they will appear (with exception of the triangle) as curved lines.

Problem 6—Find the curve cut by a plane passing through the apex of the cone shown in figure 7.13.

Let *ABC* be the elevation of the cone and *MS* the cutting plane passing through the apex. project point *D* down to the plan and parallel with the axis, cutting the base of the cone at *D'* and *E'* and obtaining *D'E'*, the base of the developed surface. With *D* as the center and *DA* as the radius (which is equal to the element of the cone-swing *A*), project a base line down to *A'*. Join *A'* with *D'* and *E*. Then, *A'D'E'* is the developed surface or triangle cut by plane *MS* with the cone.

Problem 7—Find the surface cut by a plane passing through a cone parallel with its base, as shown in figure 7.14.

This may be found by projecting over to the plan. Where

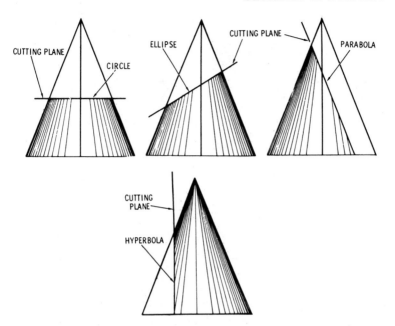

Fig. 7.12. Conic sections obtained by cutting a cone with a plane.

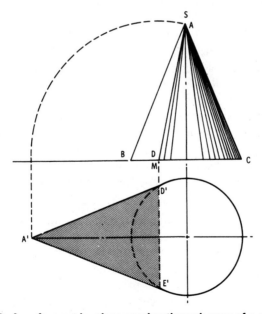

Fig. 7.13. A surface cut by plane passing through apex of a cone triangle.

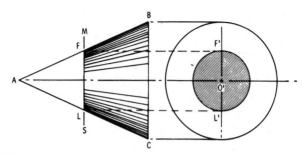

Fig. 7.14. A surface cut by plane passing through a cone parallel with its base.

MS cuts the element *AB* (as at *F*), project over to the axis of the plan and obtain point F'. Similarly, point *L'* may be found. These points are equidistant from the center *Q*. Hence, with the radius equal to *O'F'* and *O'L'*, describe a circle which is the curve cut by plane *MS* when parallel with the base of the cone.

Problem 8—Find the curve cut by a plane passing through a cone inclined to its axis, as shown in figure 7.15.

In figure 7.15, let the plane cut the elements *OA* and *OB* of the curve at *M* and *S*, respectively. Project *S* down to *s* in the plan, giving one point on the curve. With *S* as the center, swing *M* around and project down to *m'* in the plan, giving a second point, *m's'*, on the curve. This is the major axis of the curve. To find the minor axis of the curve, bisect *MS* at *R* and swing *R* around and draw radius 3. Describe an arc with radius 3, with *O'* as the center. Where this cuts the projection of *R* at *r*, project over to the plan, intersecting the vertical plan projection of *R* at *r'*. Projection of *O''r'* is half the minor axis. To find the projection of any other point, such as *L* or *F*, proceed in a similar manner as indicated, obtaining *l'* or *f*. The curve joining these points and symmetrical points below the major axis is an *ellipse*.

Problem 9—Find the curve cut by a plane passing through a cone parallel with an element of the cone, as shown in figure 7.16.

Let the plane *MS* cut element *AB* at *L* and at the base *F*. Project *F* down to the plan, cutting the base at *F'* and *F''*, which are two points in the curve. With *F* as the center and *LF* as the radius, swing point *L* around and project down to the axis of the plan, obtaining point *L'* in the curve. Now, any other point such as *R* may be obtained as follows: swing *R* around with *F* as the

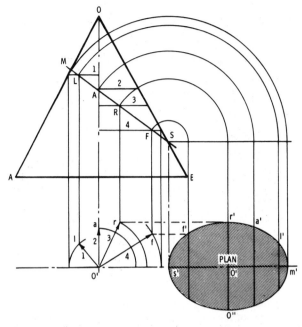

Fig. 7.15. A surface cut by a plane passing through a cone inclined to its axis.

center and project down to the plan with line *HG*. Describe an arc in the plan with a radius equal to the radius *rr'* of the cone at the elevation of point *R* and where it cuts the projection of *R* at *R'*. Project *R'* over to line *HG* and obtain point *R''*, which is a point in the curve. Other points may be obtained in a similar manner. The curve is traced through points *F'*, *R''*, *L'*, and similar points on the other side of the axis, ending at *F''*. Such a curve is called a *parabola*.

Problem 10—Find the curve cut by a plane passing through a cone parallel with the perpendicular axis of the cone, as shown in figure 7.17.

Let plane *MS* cut element *AC* at *L* and the base of *F*. Project *F* down to the plan, cutting the base at *F'* and *F''*, which are two points in the curve. With *F* as the center and *FL* as the radius, swing point *L* around and project down to the axis of the plan, obtaining point *L'* in the curve. Now, any other point (as *R*) may be obtained as follows: swing *R* around with *F* as the center and project down to the plan with line *HG*. Describe a circle in

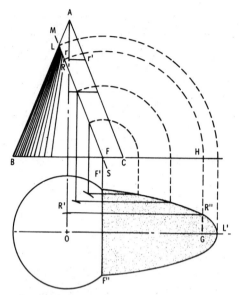

Fig. 7.16. A surface cut by plane passing through a cone parallel with an element of the cone.

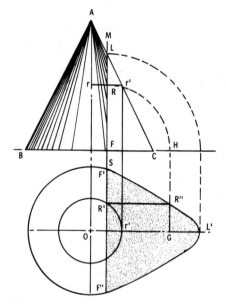

Fig. 7.17. A surface cut by a plane passing through a cone parallel with the axis of the cone.

the plan with radius *rr'* equal to the radius of the cone at elevation of point *R* and where this circle cuts the projection of *R* at *R'*, project over to line *HG* and obtain point *R''*, which is a point in the curve. Other points may be obtained in a similar manner. The curve is traced through points *F'*, *R''*, *L'*, and similar points on the other side of the axis, ending at *F''*. Such a curve is called a *hyperbola*.

It is important to remember that many of the points in this chapter are very useful, especially in sheet metal work. Moreover, a basis is established for understanding the next chapter, which covers orthographic drawings.

CHAPTER 8

Orthographic Drawings

Preliminaries leading up to a discussion of orthographic projection, which is used in blueprint drawing, have been covered in the preceding chapters. Orthographic projection is a method of representing the exact shape of an object by means of two or more views, generally at right angles to each other. These are obtained by dropping perpendiculars from the object to the plan. A thorough understanding of the preceding chapter will make the understanding of this chapter easier.

At first, orthographic drawings may appear hard to understand. We are usually able to see more than one side of an object. In other words, we receive a pictorial view of it. Pictorial drawings are not practical in working drawings, however, because the size of the drawing would be a hindrance. With an orthographic projection, one immediately obtains a pictorial image. Both pictorial drawings and orthographic projections have definite advantages, but the latter is by far the most universally used.

The experienced mechanic or drafter will tell you that, in orthographic projection, his mind is not diverted by numerous lines; instead the lines guide him directly to a picture of the object that they represent. As pointed out in the last chapter, pictorial drawings are based on *radial* lines of sight—that is, radial lines of projection—but orthographic drawings depend on *parallel* lines of sight. Instead of considering the eye at an infinite distance for parallel lines of sight, it is simpler to assume that the position of the eye is changed for each line, so that it is directly in a perpendicular line with each point of an object viewed. This is shown in figure 8.1.

In figure 8.1, let $ABCD$ be the object. View each point by moving the eye to positions $A'B'C'D'$, so that the line of vision will, in each case, be through the point and perpendicular to the projection plane, striking the latter in points $A''B''C''D''$. If these

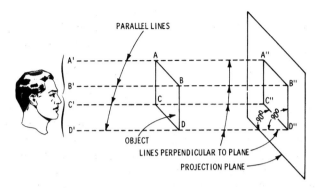

Fig. 8.1. A diagram illustrating the effect of parallel lines of vision.

points were marked on the plane and the object moved perpendicular to the plane, then the points $ABCD$ of the object would coincide with points $A''B''C''D''$ marked on the plane. Compare this with radial rays, in figure 8.2, with the eye in one position.

What is the radial projection effect in figure 8.2 when the object is between the eye and the projection plane?

Answer: The outline of the object on the projection plane is *larger* than the object, as shown in figure 8.2.

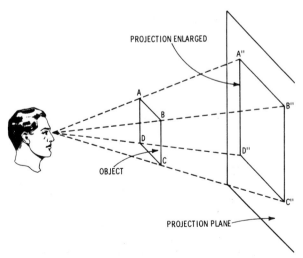

Fig. 8.2. A diagram illustrating the effect of radial projections with an object between the eye and the projection plane.

What is the effect when the projection plane is between the eye and the object?

Answer: The outline of the object on the projection plane is *smaller* than the object, as shown in figure 8.3.

With parallel projection, what is the effect of changing the relative position of the object and the eye?

Answer: The projection of the object on the plane is always the same size as the object, as shown in figure 8.4. Accordingly, the drafter can scale the projections of an object and place the dimensions on the drawing since the perpendicular projections are always the same.

When is the projection not true to size?

Answer: The projection is not true to size when the side of the object being projected is not parallel with the plane in which the object lies.

How is this overcome?

Answer: The projection can be made by projecting on an *auxiliary* plane parallel with the side of the object being projected, which is explained later in this book.

ORTHOGRAPHIC PROJECTION VIEWS

In the orthographic method, several views are required to completely show an object. The number required depends upon the shape of the object and the position of views relative to each other.

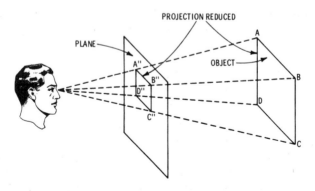

Fig. 8.3. A diagram illustrating the effect of radial projection with projecting plane between eye and object.

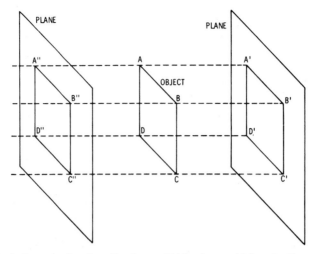

Fig. 8.4. A diagram showing that in parallel (orthographic) projection, the projection is always the same size on the side of the object projected.

In explaining orthographic projection views in this section, the illustrations will show *second-* and *third-* angle projection. For simplicity, self-defining names are given here to the various views, although different names are preferably used.

1. Top view is known as a *plan*
2. Bottom view is known as a *bottom plan*
3. Front view is known as a *longitudinal elevation*
4. Left-side view is known as a *left-end elevation*
5. Right-side view is known as a *right-end elevation*

From this it is seen that any view is either a *plan* or an *elevation*.

What is a plan?
   Answer: A plan is a horizontal view of an object.

What is an elevation?
   Answer: An elevation is a vertical view of an object.

PROJECTED VIEWS

The various views are projected on imaginary projection planes, similar to the projection of a picture on a camera plate or ground glass. The difference is that the orthographic projection lines are parallel and perpendicular to the surface instead of radial as in photography. To illustrate the first or front view, place a clear pane of glass in front of the object with the glass parallel to the surface of the object being projected. Figure 8.5 shows a rectangular object with an inclined top resting on the horizontal glass plane, placing it in the second quadrant. In front of the object is the pane of glass marked V that represents a vertical plane.

When an observer looks from a considerable distance through the glass directly at the front of the object, he will see only one side, in this instance the side marked *ABCD*. The rays of light falling upon the object are reflected into the eyes of the observer, and in this manner he *sees* the object. The pane of glass (vertical plane) is placed so that the rays of light from the object will pass through the glass in *straight parallel lines* to the eyes of the observer. In reality, the lines are not parallel but are radial. However, to meet the orthographic projection requirements they are assumed to be parallel, which they would be in this instance if the object were viewed at an *infinite distance.*

The rays of light (orthographic projection lines) from points *ABCD* of the object pass through the glass at points *a*, *b*, *c*, and *d*.

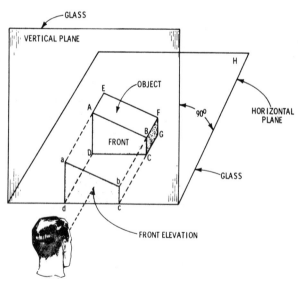

Fig. 8.5. A rectangular object resting on a horizontal plane and facing a vertical plane.

If these points are connected by lines, a view of the object as seen from the front, called a *front elevation*, is obtained.

For the particular object shown in figure 8.5, what may be said of the front elevation?

Answer: It is identical as to shape and size with the front side *ABCD* of the object; that is $ab = AB$; $bc = BC$, etc.; angle dab = angle *DAB*; angle abc = angle *ABC*, etc.

TOP VIEW OR PLAN

For this view, place a pane of glass in a horizontal position above the object that is resting on the horizontal plane, as shown in figure 8.6. Now, looking at the object directly from above, the rays of light (orthographic projection lines) from corners *AEFB* of the top pass through the glass at points *aefb*. If these points *aefb* are connected by lines, a view of the object as seen from the top, called the *top view* (preferably *plan*), is obtained.

In what quadrant is the object in figure 8.6 viewed?

Answer: It may be in either the third or fourth quadrants, depending upon the placement of the vertical plane (not shown here.)

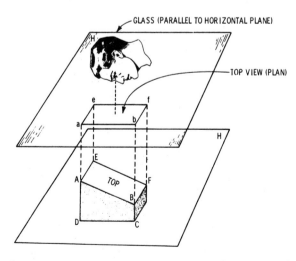

Fig. 8.6. Projection of the top of an object to obtain top view.

RIGHT-END VIEW (ELEVATION)

A pane of glass is placed to the right of the object in a vertical position and parallel to the right side *BFGC* of the object, as shown in figure 8.7. The pane of glass is marked *P*, which means *profile plane*, or plane from a side projection. Looking at the object directly from the right side (as in position *S*), the rays of light (orthographic projection lines) from corners *BFGC* of the upper-left-hand side (from points *AE*) pass through the glass at points *bfgc* and *ae*. If these points are connected by lines, a view of the object as seen from the right side, called the *right-side view* (preferably *right-end elevation*), is obtained.

What is the peculiarity of the elevation?

Answer: The shape of the object is such that the entire visible surface does not lie in a plane parallel to the projection plane. The points *A* and *E*, though located at the other end of the object, are visible and accordingly form part of the right-end view.

Does *aefb* show the top in its true size?

Answer: No, because it is projected obliquely instead of at 90°.

How does oblique projection affect size?

Answer: It makes an object appear smaller than its real size.

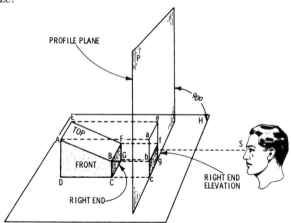

Fig. 8.7. Projection of the right end of an object illustrating right profile plane.

LEFT-END VIEW (ELEVATION)

With a pane of glass shifted to the left side of the object, as in figure 8.8, and the object viewed directly from the *left* side (as position S), the rays of light (orthographic projection lines) from corners *ADHE* of the left side pass through the glass at points *adhe*. If lines connecting these points are drawn on the glass, a left-side view of the object is obtained.

Does this complete the view?
    Answer: No. The edge *FB* at the other end is invisible.

How is it shown?
    Answer: By a dotted line connecting *f* and *b* projected from *F* and *B*. The completed drawing is then called a *left-side view* (preferably *left-end elevation*).

PROJECTION OF OBLIQUE SURFACES

Some irregularly shaped objects will have a side that will not be parallel or horizontal with any of the projection planes.

How is a true projection obtained of an oblique surface?
    Answer: By projecting on an oblique or auxiliary plane, such as shown in figure 8.9. Here the horizontal and vertical

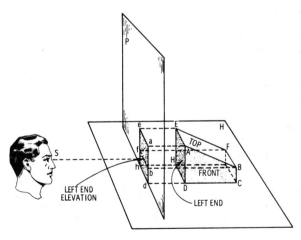

Fig. 8.8. Projection of the left end of an object illustrating left profile plane.

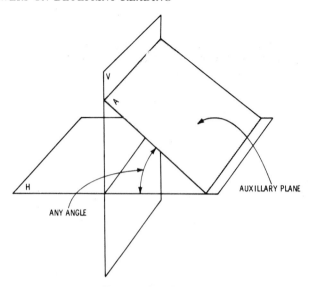

Fig. 8.9. Auxiliary plane.

planes are marked *H* and *V*. The shaded plane marked *A* is laid out at the same angle as that of the surface to be projected, so that the plane will be parallel with the surface.

Why should the auxiliary plane be parallel with the surface to be projected?

Answer: A perpendicular 90° projection is necessary to prevent distortion, giving a projection of the object in its true size and shape.

In figure 8.10, the auxiliary plane *A* is shown at such an angle that it is parallel with the oblique surface *AEFB* of the object. Now, when the object is viewed through glass directly in line, as position *S*, the rays of light (orthographic projection line) from corners *AEFB* of the object project perpendicular to the surface *AEFB* and pass through the glass at points *aefb*. If lines connecting these points are drawn on the glass, a view of the *oblique top* of the object is obtained, which is identical in size with the surface of the object projected.

ARRANGEMENT OF VIEWS

Since several orthographic views are necessary to completely show an object, the order in which these are laid out on a flat sheet of

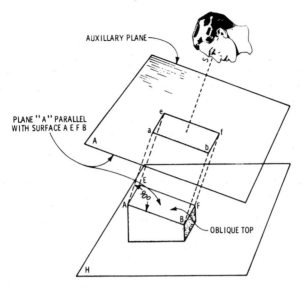

AUXILLARY PLANE

PLANE "A" PARALLEL
WITH SURFACE A E F B

OBLIQUE TOP

Fig. 8.10. Projection of inclined side of an object illustrating application of auxiliary plane to obtain true projection size.

paper is important, because improper grouping will render the drawing worthless. It is difficult to represent a solid object on a flat piece of paper. It can be shown only in part by *cabinet* or *isometric* drawings. Hence, orthographic drawings are employed to show an object fully. As shown below, the order of grouping will depend on the quadrant or angle of projection used.

First, the arrangement universally used in this country is *third-angle projection*. For illustrating a rectangular object such as a cigar box, it is shown in cabinet projection (figure 8.11A) and then *unfolded* so that all views come into the plane of the paper, as shown in figures 8.11B and 8.12. Imagine the various sides hinged so they could open as shown in figure 8.11B. Here they are shown partially open, and the view shows pictorially *how* they open. When the various sides have opened fully so that they are in the same plane

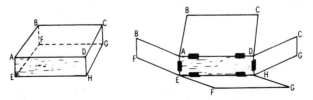

Fig. 8.11. An orthographic drawing of a cigar box.

93

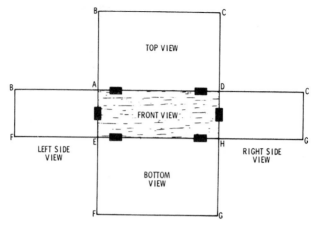

Fig. 8.12. Five views of an object as drawn in orthographic projection.

as that of the front view, the assembly will look like the illustration shown in figure 8.12. This represents the order of the views universally used.

Starting with the front view (figure 8.12), the sequence of views is as follows:

1. Top view is *above* the front view
2. Bottom view is *below* the front view
3. Right-side view is at the *right*
4. Left-side view is at the *left*

This arrangement of views is logical and is accordingly easy to remember. For such a simple object as the cigar box, two views would suffice, but the other views shown here bring out the arrangement, or order of views. This arrangement corresponds to *third-angle* (quadrant) projection.

QUADRANTS OR ANGLES OF PROJECTION

Since, as shown in the foregoing illustrations, the vertical and horizontal planes intersect, there will be four quadrants designated as angles (figure 8.13). Evidently the object to be projected may be placed in any of these quadrants.

How is the object viewed in any quadrant?

Answer: So that the line of sight is perpendicular to the

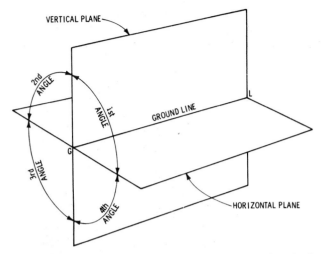

Fig. 8.13. Intersecting vertical and horizontal planes illustrating the four quadrants or angles as they are usually called.

projection planes. In the first-angle projection, the object is always viewed as if directly *before* the eye, or directly *below* the eye, as shown in figure 8.14.

With respect to how the object is viewed, what is the important difference in the first and third quadrants?

Answer: The object is viewed *directly* in the first quadrant, and *through glass* in the third quadrant. The object is between the eye and the projection planes in the first quadrant, and the *glass* (projection planes) is between the eye and the object in the third quadrant.

What is the effect of the change in the relative position of the eye, the object, and the projection planes?

Answer: It changes the order or arrangement of the views.

How are the lines of an object projected in first- and third-angle projection?

Answer: In first-angle projection, they are projected *away* from the eye to the projection planes beyond the object. In third-angle projection, they are projected *toward* the eye through the glass (projection planes) and traced thereon.

This is clearly shown in figure 8.15 for the projections on the vertical plane. The same conditions will be obtained for the projection on the horizontal plane.

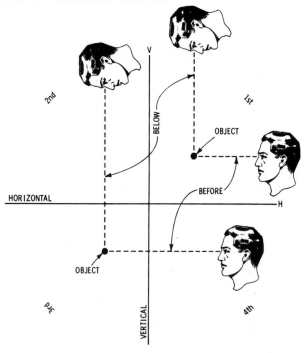

Fig. 8.14. End view of vertical and horizontal planes illustrating how an object is viewed.

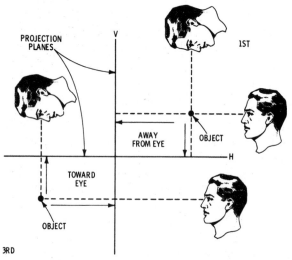

Fig. 8.15. Method of viewing an object in first and third quadrant.

Why does the top plan come below the elevation in first-angle projection?

Answer: Because the projection on the horizontal plane (plan) is revolved *downward* to bring it into the vertical plane.

Note in figure 8.16A that the plan projected on the horizontal plane is the top of the object, because the object is between the eye and the plane. Accordingly, in such cases the projection will be that side nearest the eye. The operation of revolving the plane downward is plainly shown in figure 8.16B, which positions the top plan view *below* the elevation.

Why does the top plan view come *above* the elevation in third-angle elevation?

Answer: Because the projection on the horizontal plane is a plan of the top, and this is revolved *upward* to bring it into the vertical.

Thus, in figure 8.17A, a cylindrical object is shown in the third quadrant, with its projections on the *H* and V planes. In viewing the object with the planes between the eye and the object (through glass), it is clearly seen in figure 8.17A that the top of the object, which is the end nearest the *H* plane, is also the one projected on the *H* plane. Similarly, the elevation is

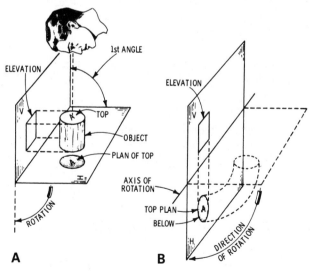

Fig. 8.16. First-angle projection of an object illustrating results obtained by this method.

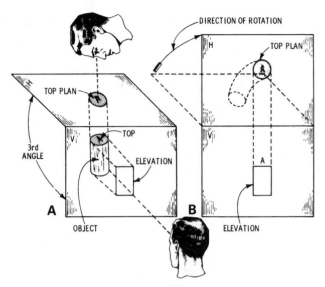

Fig. 8.17. Third-angle projection of object. Plan comes on top and is not inverted.

projected on the V plane, as shown. Now, in revolving the H plane into the vertical, the rotation is upward, as shown by the arrow in figure 8.17B. From figures 8.17A and 17B, it is evident that in third-angle projection, the top plane comes *above* the elevation.

END VIEWS IN FIRST-ANGLE PROJECTION

The differences to be noted in first- and third-angle projection is the reversal in the positions of the various views. This has just been explained in the case of *plan* and *elevation*, and it remains to show the reason for reversal of the end views in first-angle projection.

1. Right elevation comes on the *left* side
2. Left elevation comes on the *right* side

Consider the projection in the first angle where the object is always placed between the eye and the glass (projection plane). Figure 8.18A shows a cylindrical object as projected on the vertical and left *profile* planes. Carefully note that the profile *plane* is at 90° to the vertical plane and at 90° to the horizontal plane.

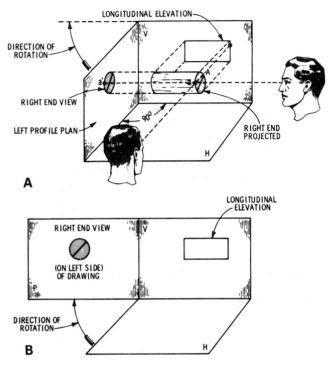

Fig. 8.18. First-angle projection of cylindrical object illustrating why right end view comes on left side of drawing.

How does line of sight project in first-angle projecation of end views?

Answer: It is projected away from the eye to the end of the object nearest the eye, and thence to the profile plane.

Thus, in figure 8.18A, if the observer views point *a* of the object and the line of vision is continued straight to the profile plane, it would locate point *a*. Clearly, then, the projection on the *left* profile plane must be the projection of the *right end* A of the object nearest the observer and remote from the left profile plan. Accordingly, in first-angle projection, the *right end* view appears on the drawing at the *left* side.

Why does the *right*-end view appear on the *left* side of the elevation?

Answer: If the profile plane is rotated 90° in the direction shown in figure 8.18A, so that it lies in the vertical plane, the right-end view is rotated around to the left side, as shown in figure 8.18B.

Why does the left-end view appear on the right side of the elevation in first-angle projection?

Answer: Similarly. as for the right-end view, it is because of the observer's position and the direction of rotation of the profile plane. Thus, in figure 8.17A, the line of sight is projected to that end of the object nearest the observer, thence to the profile plane.

Accordingly, if in figure 8.19A the observer viewed point *b* of the object, and the line of vision is continued straight to the right profile plane, it would locate point *b'* thereon. Then the projection on the right profile plane must be the projection of the left end *B* of the object nearest the observer and remote

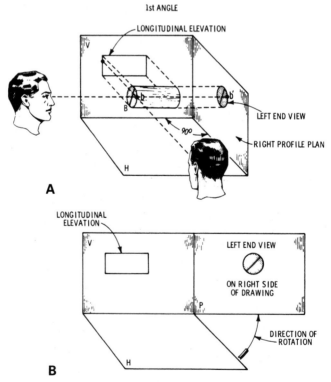

Fig. 8.19. First-angle projection of cylindrical object illustrating why left end view comes on right side of drawing.

from the right profile plane. In the first-angle projection, the left-end view appears on the drawing at the right side. This is shown in figure 8.19B. Note the rotation of the profile plane, as indicated by the arrow, that is, 90° to bring it in the vertical plane.

END VIEWS IN THIRD-ANGLE PROJECTION

In orthographic drawings using third-angle projection, end views and, in fact, all of the views appear in their logically correct place with respect to the end views. In *third-angle* projection

1. Right elevation comes on the *right* side
2. Left elevation comes on the *left* side

Considering third-angle projection, where the object is always viewed *through glass*, the imaginary projection planes (assumed to be transparent as if made of glass) are always placed between the eye and the object, as indicated in figure 8.15.

How are the lines of an object projected in third-angle projection of end views?

Answer: They are projected from the object toward the eye to the glass or projection plane and traced thereon. In figure 8.20A, for a left side view, note the observer at the left and the profile plane (glass) between the observer and the object.

Why does the left-end view come on the left-hand side when the profile plane is rotated into the vertical plane?

Answer: Because it is the nearest or left-end view *B* of the object that is projected, as clearly shown in the illustrations.

Note that it is not the direction in which the profile rotates, but the particular end that is projected to the profile plane that determines how the object appears. In this instance, it is the left end view *B* that is projected, instead of the opposite or remote end *A*, as would result in first-angle projection.

How is the right-side view projected?

Answer: As shown in figure 8.21A. As for all third-angle projections, the projecting plane is between the observer and the object.

As viewed in figure 8.21A, the side *A* of the object that is

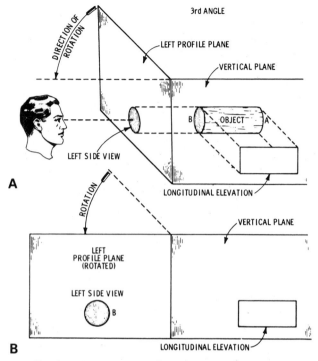

Fig. 8.20. Third-angle projection on left profile plane.

nearest the observer is projected to the intervening right profile plane. For this reason, when the profile plane is rotated into the vertical plane, the right-side view comes on the right side, as shown in figure 8.21B.

BOTTOM AND BACK VIEWS

These views, for either first- or third-angle projection, are projected according to the same basic principles as have been applied at such great length to the other views.

SECOND- AND FOURTH-ANGLE PROJECTION

In a study of these projection quadrants, it will be found that

1. Plan and elevation lie *above* the ground line for the second quadrant.

2. Plan and elevation lie *below* the ground line for the fourth quadrant.

Why are the second and fourth quadrants seldom used in practice?

Answer: Because, in revolving the *H* plane into the vertical, the plan views become *inverted.*

This is evidently objectionable, and since otherwise the drawings are identical with those of the first and third quadrants, there is no reason for using the second and fourth.

Show why in second-angle projection the plan is inverted and below the elevation.

Answer: In figure 8.22A, since the projection on the vertical plane is of the end *A* (nearest the plane), evidently when the horizontal plane is rotated upward into the vertical, as in figure 8.22B, the plan will become above the ground line, as shown. Relative position of plan and elevation depends upon the location of the object in space.

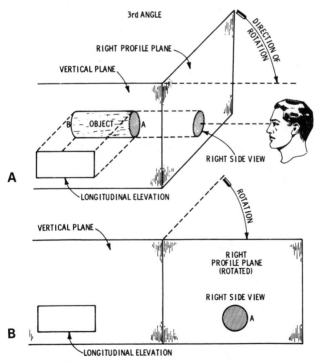

Fig. 8.21. Third-angle projection on right profile plane.

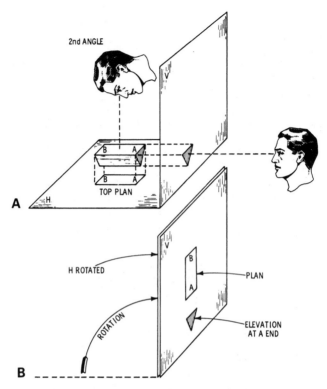

Fig. 8.22. Second-angle projection of a wedge-shaped object illustrating that plan comes below elevation and is inverted.

Show why in fourth-angle projection the plan is below the ground line.

Answer: The downward rotation of the *H* plane puts the plan below the ground line.

Using a pyramid for illustration, the projections are shown in figure 8.23A. Since the object is viewed direct for the elevation, the projection of the elevation will be of the side remote from the *V* plane, as indicated by the edge *OA*. On rotation, the plan comes above the elevation. By comparing figures 8.23A and 8.23B, it can be seen how the plan becomes inverted.

SECTIONS

Sometimes an object to be drawn is of such shape that it cannot be clearly represented by a plan and elevation. In such instances, the

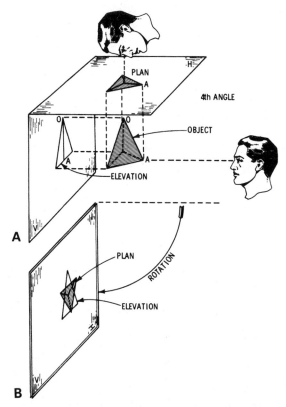

Fig. 8.23. Fourth-angle projection of a pyramid illustrating that both plan and elevation come below ground level.

parts that do not appear properly in these drawings are better represented by a *section*, or by *sectional views*.

What is a section?

Answer: A section is an orthographic drawing of an object as it would appear if cut through by an intersecting plane. A section shows nothing but the part cut by the intersecting plane.

What are the three kinds of sections usually drawn?

Answer: (1) Cross section, (2) longitudinal section, and (3) oblique section, classed according to the angle of the cutting plane.

What is a cross section?

Answer: A cross section is an orthographic drawing of an

object showing that part cut by a plane at right angles to its longitudinal axis.

What is a longitudinal section?

Answer: A longitudinal section is an orthographic drawing of an object showing that part cut by a plane at an oblique angle.

What is an oblique section?

Answer: An oblique section is an orthographic drawing of an object showing that part cut by a plane at an oblique angle.

What is a sectional view?

Answer: A sectional view is an orthographic drawing in which the parts of an object cut by the intersecting plane appear part in section and part in full view. The object of this is to save time and make the drawing easier to read.

The sectional view is used to show clearly the shape and operation of an object.

How is a section through a thin part, such as the shell of a boiler, usually represented?

Answer: In solid black instead of section lines. figure 8.24 shows a *cross section* of a piston and illustrates the use of solid black for thin parts, such as piston rings. If such parts were section lined, they would not come out sharply and well defined as with the solid black section. In sections that cut through separate and adjacent parts, these parts are distinguished by drawing the section lines of the second part at a different angle than those of the first part, as shown in figure 8.25.

CROSS SECTION

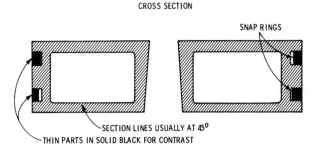

SNAP RINGS

SECTION LINES USUALLY AT 45°
THIN PARTS IN SOLID BLACK FOR CONTRAST

Fig. 8.24. Cross section of engine piston illustrating sectioning by lines and in solid black.

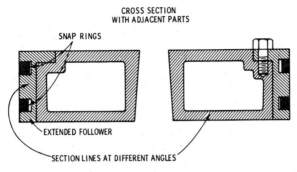

CROSS SECTION
WITH ADJACENT PARTS

SNAP RINGS

EXTENDED FOLLOWER

SECTION LINES AT DIFFERENT ANGLES

Fig. 8.25. Cross section of engine piston with extended follower (ring carrier) illustrating the method of section lining adjacent parts to show that they are not integral.

Figure 8.26 illustrates a *sectional view*. Drawings of this kind are made of objects whose construction is the same on each side of a center line or axis. Evidently, here it would be a waste of time to show everything intersected by the cutting plane in section, because section lining is tedious and requires extra time, and it is difficult to space the section lines equally. Moreover, by showing part of the object in *view*, a better idea is obtained of its appearance. Extra details can be shown in the view portion that would not appear in the section portion, for instance, the split portion of the ring.

SECTIONAL VIEW

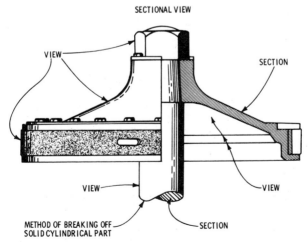

VIEW

SECTION

VIEW

VIEW

METHOD OF BREAKING OFF
SOLID CYLINDRICAL PART

SECTION

Fig. 8.26. Sectional view of a mushroom-type piston showing portions in view and in section.

DIRECTIONS OF VIEW
FOR SECTIONAL VIEWS

For an unsymmetrical object, it is important to know the direction in which the sectional view is viewed. This is indicated by arrows at the end of the line representing the cutting plane, as shown pictorially in figures 8.27A and 8.27B. The arrows AA indicate that the object is viewed in the direction of point L (toward the smaller end of the object), and the arrows BB (in the direction of R) toward the larger end. Sectional views at AA and BB would appear respectively as at L and R in figures 8.28A and 8.28B.

SECTIONS ON ZIG-ZAG INTERSECTION PLANES

A section is not always taken along one straight plane, but is often taken along zig-zag planes or even a curved intersecting surface.

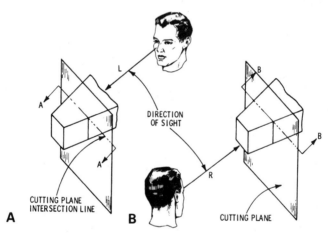

Fig. 8.27. Direction in which a sectional view is viewed is indicated by the direction of arrows AA at the ends of the cutting plane intersecting line.

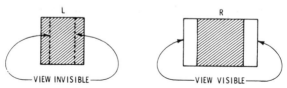

Fig. 8.28. Appearance of a sectional view with respect to the direction in which it is viewed.

The intersecting plane may be offset or become a curved surface if, by doing so, the construction can be shown to better advantage, as shown in figures 8.29A and 8.29B.

NUMBER OF VIEWS REQUIRED

The number of views depends upon the nature of the object to be represented. In any case, there should be enough orthographic projections, sections, and detail drawings to completely represent the object in all its parts. This is necessary so that the man in the shop may construct the object without loss of time in asking questions. The blueprint should give all the information required by the mechanic. In the case of a complicated object, a cabinet or isometric drawing in addition to the orthographic projections is sometimes helpful. With such an addition, the mechanic can more easily read the blueprint and avoid a misconception of the general appearance of the object to be constructed.

Usually two views, such as plan and elevation, suffice for sym-

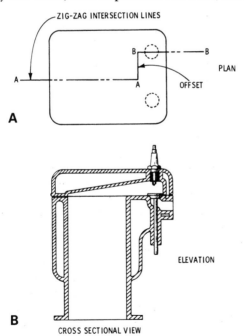

Fig. 8.29. Section of a gas engine cylinder on zig-zag intersection line AABB.

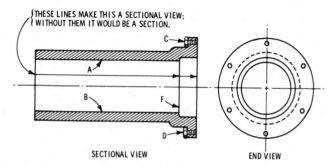

Fig. 8.30. A flanged cylinder or symmetrical object that may be completely represented by two views.

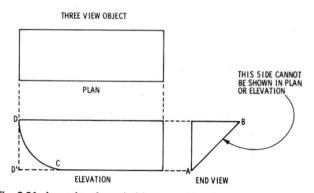

Fig. 8.31. Irregular-shaped object requiring three views.

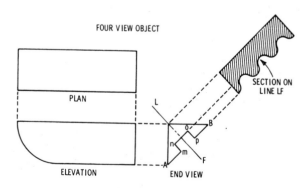

Fig. 8.32. Irregular-shaped object requiring four views.

metrical objects, but for most irregular shapes additional views are required. A flanged cylinder, for instance, may be completely represented by two views—a longitudinal sectional view, and an end view, as shown in figures 8.30A and 8.30B. To save time, an external or full view could be drawn, instead of the sectional view, by omitting the section lining and representing the inner walls *AB*, drilled holes *CD*, and edge *F*, by dotted lines.

Three-View Objects

In the case of numerous irregular shapes, more than two views are required to show the object completely. For instance, figure 8.31 illustrates an object requiring three views. Evidently the inclined back side *AB* is not visible in either the plan or elevation, and the rounded end *CD* cannot be shown in either plan or end view. If, instead of the rounded end *CD*, it was rectangular, as *CD'D* (dotted lines), the object could be completely shown by two views, plan and end view. Because of the inclined side *AB* and rounded end *CD* three views are necessary.

Four-View Objects

By slightly modifying the three-view object shown in figure 8.31, a corrugated groove could be cut in the inclined side *AB*; thereby a fourth view would be necessary to show the corrugations in their

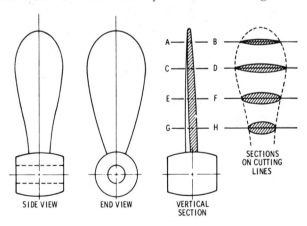

Fig. 8.33. Drawing of a propeller wheel requiring more than four views.

111

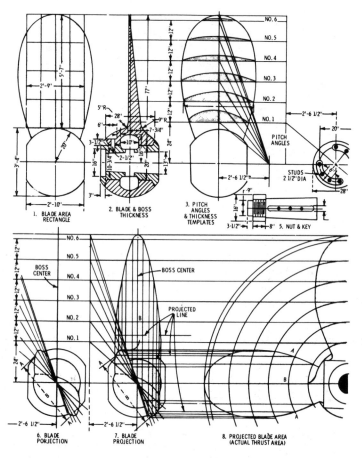

Fig. 8.34. Fully dimensioned working drawing of a built-up propeller requiring a multiplicity of views including many sections.

actual shape. In figure 8.32, the actual shape of the side *no* of the groove *mnop* is shown by a fourth view with section line *LF*.

Objects Requiring More Than Four Views

A typical example of an object requiring a multiplicity of views is a propeller for a vessel. Figure 8.33 shows some of the drawings required, including the numerous cross sections on lines *AB*, *CD*, etc. However, a complete design for a propeller wheel would appear as shown in figure 8.34.

CHAPTER 9

Working Drawings

A working drawing is usually in the form of a blueprint. It is simply an orthographic drawing, fully dimensioned, and containing data necessary for the mechanic to do the job indicated.

What is most important on a working drawing?

Answer: The dimensions. There should be a dimension for every part necessary for the mechanic to do the required work.

What should be included with the measurements of adjacent parts?

Answer: An overall measurement.

Why?

Answer: It is a check on the adjacent measurement and avoids adding the several measurements to get the overall distance. In figure 9.1, note that there are seven dimensions for adjacent parts of a crankshaft. The overall dimension is necessary for the machinist to know how long the shaft should be between ends, which is *overall.*

Is a dimension that spans several dimensions of adjacent parts necessarily an *overall* dimension?

Answer: No, in the sense that it does not cover the entire length of the object. Thus, in figure 9.2, dimension A (5″) gives the length of the threaded and unthreaded part of the bolt but not the thickness of the head. Dimension B is the actual overall dimension.

What has been put on the drawing (figure 9.2) that is not necessary?

Answer: The abbreviation "hex." Hex is the abbreviation for hexagon. It is not necessary since the view of the head is sufficient to show that the head is hexagonal.

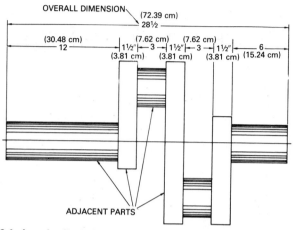

Fig. 9.1. Longitudinal view of a shaft illustrating overall dimensions.

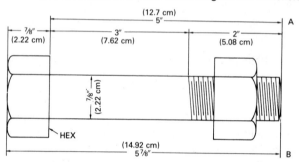

Fig. 9.2. A drawing of a bolt, illustrating dimension covering adjacent dimensions.

What important dimension has been omitted from the drawing?

Answer: The "distance between flats" of the head.

What is understood by "distance between flats"?

Answer: The distance between two opposite and parallel sides of the head or nut, as shown in figure 9.3. As explained above, the machinist must know this distance in spacing for right- and left-side milling cutters, which machines the sides of the head in pairs.

What should not be put on drawings of parts to be machined, and why?

Answer: Inch marks (″) because they are not necessary. They clutter up the drawing and obscure the dimensions; moreover, putting an inch mark after each dimension is a waste of time. Everyone will know that dimensions on a blueprint, such as a drawing of a small article like a watch, clock, or reasonably small machine part, are not in feet or yards, but in inches.

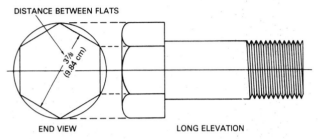

DISTANCE BETWEEN FLATS

3/8 (9.84 cm)

END VIEW　　　　LONG ELEVATION

Fig. 9.3. A method of dimensioning a bolt head or nut.

What should be placed on the drawing in special cases to prevent a mistake in omission of inch and foot marks?

Answer: The notation: *All dimensions in inches and fractions thereof unless otherwise specified.* This should go near the title of the drawing under the scale.

What should be considered in dimensioning a drawing?

Answer: The effect of locating the dimensions with ease when reading the drawing. In figure 9.4, a plan and elevation of a cylinder head are shown, with all necessary dimensions for machining. Note that all diameters are always clearer when shown on an elevation than on a plan. Accordingly, only one dimension is shown on a plan (the bolt circle); the other three diameters are on the elevation. The parts to be finished are marked *f*. In reading the plan, note that the top of the head is to be finished from *A* to *B*, the depressed part of the head from *B* to center being left rough.

Figure 9.5A shows the result of putting the dimensions for all diameters on the plan. It's a jumble of dimensions put on at various angles, tending to cause mistakes. Another criticism of figure 9.5A is that the dimensions are not put on in an orderly manner. If they must appear on a plan, a better arrangement is shown in figure 9.5B. This still leaves much to be desired and shows that numerous diameter dimensions should not be put on a plan.

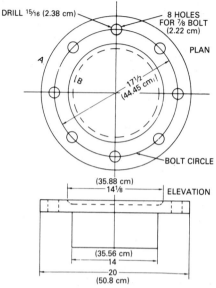

Fig. 9.4. Proper location of dimensions for easy reading of working drawing.

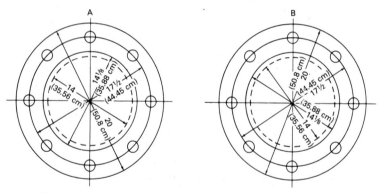

Fig. 9.5. A method of locating diameter dimension on plan.

LISTING SURFACE FINISHES

The types of finished surfaces should be indicated by a note and an arrow. The meaning of the different shop operations involved can be found by consulting a good shop handbook. Holes should be marked according to how they are formed or treated, for example, core, drill, bore, punch, ream, tap, counterbore, countersink, etc.

Surfaces should be marked to specify the shop operations that are to be performed: finish, rough, rough finish, chipped, spot faced, scraped, ground, polished, filed, etc. The kind of fit is sometimes marked as loose fit, running fit, driving fit, forced or pressed fit, shrink fit, etc. The preferred practice is to list the tolerance or limits of accuracy required, as suggested in figure 9.6A.

GEOMETRIC DIMENSIONING AND TOLERANCE

Geometric dimensioning and tolerance (GD&T) is a means of specifying engineering design and drawing requirements with respect to the actual function and relationship of part features. This is a new approach to drawings and is a technique which, when properly applied, ensures the most economical and effective production of these features. Thus GD&T can be considered both an engineering design language and a functional production and inspection technique.

The proper use of GD&T techniques will facilitate the communication of functional design requirements. Its use for nonfunctional specifications can be costly and is therefore discouraged.

Geometric dimensioning and tolerance may be used to control flatness, straightness, roundness, angularity, perpendicularity, parallelism, cylindricity, profile of any surface, profile of any line, accuracy (figure 9.6), run out, true position, concentricity, and symmetry.

A valuable reference work is *Geometric Dimensioning and Tolerance—A Working Guide*, by Lowell W. Foster of Honeywell, Inc.

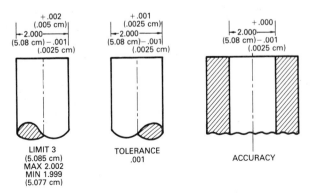

Fig. 9.6. Tolerance and limits of accuracy.

FULL VIEWS

Working drawings may be made of many objects with symmetrical forms by full views only, no sections being necessary. A familiar example of the use of full views is the shaft coupling shown in figure 9.7A. Here the object is completely shown by a longitudinal view and an end view. This is a coupling for a marine engine whose shaft is larger in diameter than the propeller shaft with which it connects. Note that the dimensions of these shafts are 3 and 2 inches, respectively. The keyway dimensions are specified instead of being dimensioned direct, owing to the smallness of the parts. The dimensions for the bolt lines are referred to the *center line* because they must be spaced the same distance on each side of the center line.

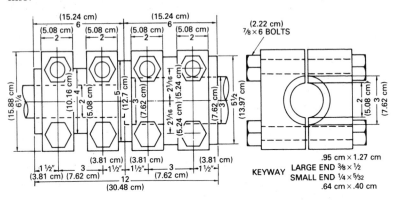

Fig. 9.7. An approved method to dimension a drawing.

What other way could working drawings be made of the shaft coupling of figure 9.7A?

    Answer: The coupling could be shown by the longitudinal view of figure 9.7A and two cross sections as shown in figure 9.8.

What is the criticism of the method shown in figure 9.8?

    Answer: Although it shows the construction clearly, it is objectionable because it is a waste of the drafter's time.

In figure 9.7, did the drafter forget to describe dotted circles showing the 4- and 5-inch round portions as dimensioned in figure 9.7B?

    Answer: No.

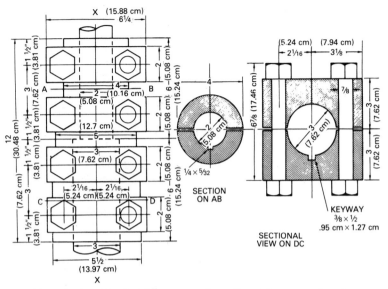

Fig. 9.8. An objectionable way to show dimensions on a drawing.

How would the patternmaker know that these sections were not
square instead of round?

Answer: Common sense. The drafter has indicated with
the large circle in figure 9.8 that the section at the end is round.
This is enough information for the patternmaker to know that
the other sections are round

**What was the drafter's chief reason in omitting the dotted circles
indicating round sections?**

Answer. To simplify the drawing and make it easily readable. By inspection, it is seen that this drawing is intended to
show only the size of the holes for shafts, bolts, and keyways.
The drawing is further simplified by giving the keyway dimensions in the legend instead of dimensioning the drawing.

**What should be noted about the arrangement of the drawings in
figure 9.8?**

Answer: The two sections are placed on the right side of
the full-view drawing instead of being "strung" on the *XX* axis
where they really belong.

Is this permissible and good practice?

119

Answer: Yes, when it is expedient to make a special arrangement of the views. In the first place, all the dimensions are vertical, which make them easy to read; moreover, the blueprint is easily handled as compared with a longer drawing.

SECTIONS AND SECTIONAL VIEWS

Sections and sectional views are necessary in many drawings to bring out and fully dimension the object. Usually one section or sectional view is sufficient, but for some irregular objects several may be required. Frequently, one sectional view is used in place of a full view and a section. Sometimes an object is of such symmetrical outline it can be completely shown with one cross section, in fact, with only one half of the cross section.

An example of showing complete representation with only a half cross-sectional view is the jacketed engine cylinder, as shown in figure 9.9. This drawing illustrates several things. First, all the horizontal dimensions are marked *D*, which means *diameter*. Some of these *D* horizontal dimensions are continued in line with separate dimensions, such as 3/8″ thickness of cylinder wall, 1/4″ minimum thickness of liner, etc. About the only thing not indicated is the number of jacket outlet passages at *A* and *B*. However, a legend may be added giving the number of air passages and number of studs.

Mention something open to criticism in figure 9.9.

Answer: The incorrect method shown of making fractions. As stated above, the division line of a fraction should be horizontal to the text instead of inclined. However, on this particular drawing, the fractions are made with such care that there is little possibility of making a mistake.

Sections for Irregular Objects

Because of their shape, some objects require more than one section for complete showing. For instance, figure 9.10 is an example in which two sectional views are necessary. As an aid to blueprint reading, the various views should be labeled, as in figure 9.10, bottom plan, section on *AB*, section on *CD*, etc., which leaves nothing to the imagination. With all dimensions put on, the views

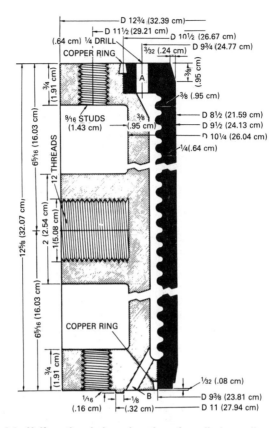

Fig. 9.9. Half-sectional view elevation of a cylinder wall.

become a *working drawing* from which the machinist obtains all the necessary information to machine the object.

Ribs in Plane View

On some objects, such as pulleys having radial ribs or spokes, it is necessary to take a section on a plane through the arm in order to show the shape of the spokes as well as the rim. When a rib comes on the plane of a section, it is desirable to distinguish between the section and the rib.

How is the distinction made?

    Answer: By using fine section lines for the section and coarse lines for the rib, as shown in figure 9.11.

121

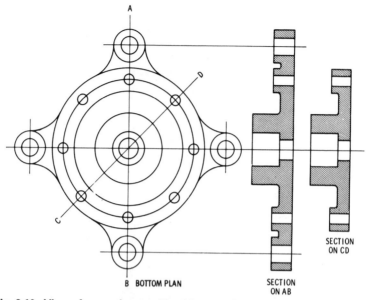

Fig. 9.10. View of an engine combined lower cylinder head, illustrating an object requiring full view and two sections.

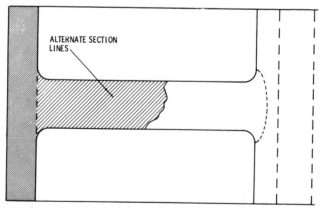

Fig. 9.11. Drawings of a pulley illustrating fine and coarse section lining, on sections with ribs in plane section.

What is the practical method of drawing the section lines?

Answer: By omitting alternate lines on the rib section. This method gives a pleasing appearance to the drawing, and the dotted contour lines clearly indicate the shape and thickness of the rim section.

How may this alternate sectioning be avoided?

Answer: By taking a section on a plane between the ribs and showing a separate or detail cross section of the ribs. Thus, in figure 9.12, the full view of the wheel is drawn first, and then the sectional view taken on *AB* midway between the ribs is drawn. This eliminates having the rib section joined onto the section of the rim and gives a better idea of the wheel. The cross shape of the ribs is clearly shown by the section taken on *CD*. Note that the three views are each labeled, which they should be for quick and easy reading.

What would be a better location for the section on *CD*?

Answer: Instead of making it a separate drawing, place it across one of the arms, as at *EF* in the full-view drawing in figure 9.12.

What help is this in blueprint reading?

Answer: The mechanic, after looking at the separate sectional view, does not have to locate *CD* in the full view. But, when the section is placed across the arms, as at *EF*, the location of the section is seen at once.

How should the principal dimensions (diameters) be put on the flywheel?

Answer: On the sectional view rather than the full view.

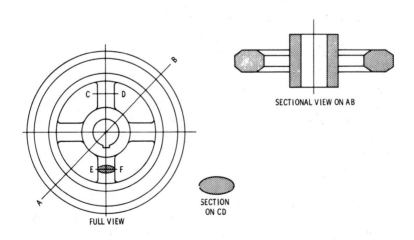

SECTIONAL VIEW ON AB

SECTION ON CD

FULL VIEW

Fig. 9.12. Plan and elevation of a Shipman flywheel.

Why?

Answer: To avoid dimensions at various angles and crowding. Evidently, when placed as shown in figure 9.13A, they are easily read; figure 9.13B shows an objectionable arrangement, the effect being virtually a mess of dimensions.

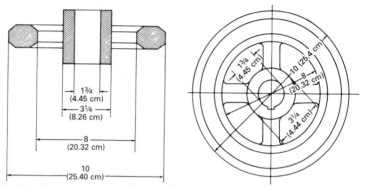

Fig. 9.13. Approved and objectionable methods of dimensioning the Shipman flywheel shown in figure 9.13.

Objects Suitable for Sectional Views

Some machine details are specially adapted for representation by sectional views, such as drawings of objects that include bolts, studs, pistons, piston rods, and other cylindrical parts. The sectional view compared with a section is more pictorial, shows the object more plainly, and saves time in drawing. A typical object for a sectional view is the familiar stuffing box. As an example of this, figure 9.14 shows a marine stuffing box for the propeller shaft of a small boat. It is fully shown by a sectional view and a right-end elevation. Considering the sectional view in figure 9.14, the center line divides the drawing into a half full view (below center line) and the sectional view proper above the center line. Here the upper half of the stuffing box and gland is in section, the box being section lined and the packing gland in solid black. The occasional use of solid black instead of section lines makes a drawing snappy and easier to read and the box and gland are better defined than if both were section lined.

What is sometimes done to make the drawing more pictorial on a sectional view, such as shown in figure 9.14?

Answer: The cylindrical parts are shaded.

An example of such treatment is shown in figure 9.15, representing a special feed pump designed to run at high speed and pressure without hammer or water knock. Note the shading on the plunger, part of which is in cross section. The stuffing box is of the screw type, the packing gland and ball retainer being in solid black section. Note how plain and sharp they appear in the drawing due to the contrast between the solid black and section lines. In the plan, note that the hexagon part of the stuffing box screw cap is dimensioned *between flats* so the machinist will know the spacing for the *right* and *left* milling side cutters. Also notice the method of putting the 1⅜″ dimension of the well onto the drawing.

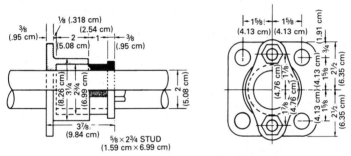

Fig. 9.14. Inside marine stuffing box.

Dotted Lines

To easily read a blueprint, it should have no more than the necessary lines to fully represent the object. This relates especially to dotted lines, whether projection or hidden-part lines. How much of the hidden part behind the plane of a section should be represented must be determined for each particular case. Often in a sectional view, only the outline of the sectional surfaces and the full lines will appear. Except in special cases, this is all that is necessary, and filling up the drawing with dotted lines will only make it difficult to read and more difficult for the drafter to find room for the dimensions.

ASSEMBLY DRAWINGS

There are two kinds of working drawings:

1. Assembly drawings

2. Detail drawings

What is an assembly drawing?

Answer: An assembly drawing is a drawing of a multipart object showing it as a whole, that is, all its parts assembled in their proper positions.

What is the object of a multipart drawing?

Answer: It shows the mechanic how to assemble the various parts that make up the object.

What is important on an assembly drawing, and why?

Answer: Complete overall dimensions to indicate how much space will be required in installation.

What drawings in this chapter are assembly drawings?

Answer: Figures 9.2, 9.7, 9.13, and 9.16.

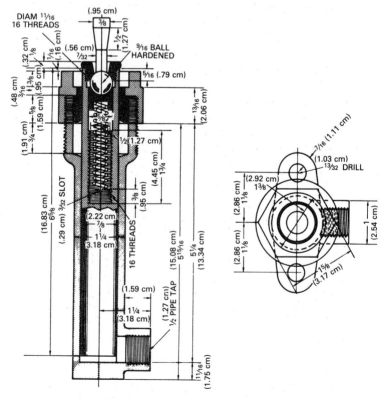

Fig. 9.15. Illustrating a spring cushion feed pump.

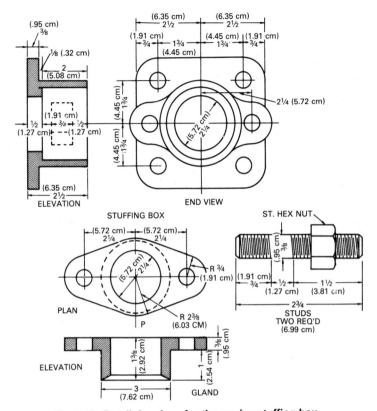

Fig. 9.16. Detail drawings for the marine stuffing box.

Are all dimensions put on assembly drawings?

Answer: Not always; sometimes only the overall dimensions.

What detail drawings would be made for the stuffing box shown in the assembly drawing in figure 9.14?

Answer: There should be separate drawings—detail drawings for the stuffing box casting, the gland, and the studs. Figure 9.16 shows these details as they would be drawn and arranged on one sheet.

What are the advantages of detail drawings?

Answer: Each part is shown more clearly and there is plenty of room for all the dimensions. Moreover, if each part

127

were given to different machinists, each man would have a separate drawing of the part on which he is working.

When the parts are given to different machinists, how are the drawings made?

Answer: They are printed on separate blueprints instead of all together on one blueprint.

CHAPTER 10

Surfaces

In pattern cutting or sheet metal work, surfaces are divided into two general classes known as *elementary* and *warped.*

What is an elementary surface?

Answer: An elementary surface is a surface in which a straight edge may be placed in continuous contact with the surface *abcdef*, as shown in figure 10.1.

What is the line of contact of the straight edge called?

Answer: An element of the surface.

What are consecutive elements of the surface?

Answer: Elements that lie infinitely close to each other.

What are the characteristics of elementary and warped developments?

Answer: An object with an elementary surface may be developed accurately, but one with a warped surface only approximately.

Objects having elementary surfaces may be formed by simply folding or rolling the metal pattern. If the object has a warped surface, the metal pattern must undergo the operation of raising or bumping to bring the pattern to the true shape of the object when folded or rolled.

Name two kinds of elementary surfaces.

Answer: Plane and curved, as shown in figure 10.2A.

What is a plane surface?

Answer: A plane surface is a surface in which elements may be drawn in any direction, as in figure 10.2A.

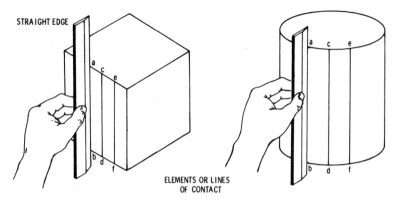

Fig. 10.1. Plane and curved elementary surfaces.

What is a curved surface?

Answer: A curved surface is a surface in which no three consecutive surface elements lie in the same plane, as in figure 10.2B.

In the figure, let *ab*, *cd*, and *ef* be three consecutive elements. If a plane, *LARF*, passes through *ab* and *ef*, the intervening element *cd* will not lie in this plane. The curved surface shown in figure 10.2B is a cylindrical or curved surface having parallel elements, as distinguished from another class of curved surfaces that does not have parallel elements, as for instance a conical surface.

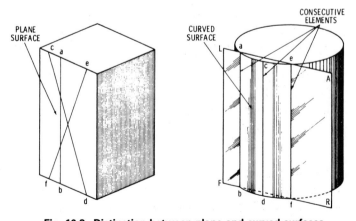

Fig. 10.2. Distinction between plane and curved surfaces.

130

What is a cylindrical surface?

Answer: A cylindrical surface is a surface having parallel elements. No three consecutive elements lie in the same plane. Figure 10.3A is a cylindrical surface.

What is a conical surface?

Answer: A conical surface is a surface having radial elements meeting in a common point. No three consecutive elements lie in the same plane. Figure 10.3B is a conical surface.

What is a warped surface?

Answer: A warped surface is a surface in which a straight edge may be placed in contact only at a point, for instance, the sphere shown in figure 10.4.

What is the characteristic of patterns for warped surfaces?

Answer: They can be cut only to the approximate shape (figure 10.5).

The pattern *LFR*, for a section of a warped surface, must be warped or hammered to the shape *lfr*, so that its surface will coincide with the warped surface. The figure clearly shows the shape of the pattern before and after warping. Note that *lf* is brought in from *L'F'* and *r* and lowered from *R'*.

How are objects having elementary surface developed?

Answer: By the methods of (1) parallel lines, (2) radial lines, or (3) triangulation.

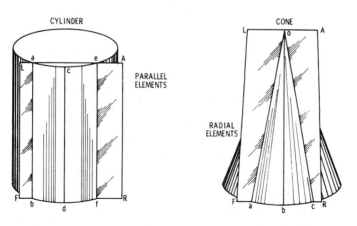

Fig. 10.3. Distinction between cylindrical and conical surfaces.

131

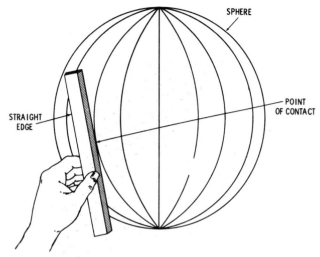

Fig. 10.4. A warped surface or surface in which a straight edge can be placed in contact only at one point.

PARALLEL LINES

In any development, a plan and elevation of the object is drawn first, and from these views the developed shape of the object is obtained by laying off what is called a *stretch-out*.

What is a stretch-out?

> *Answer:* A stretch-out is the outline of the object unfolded and laid flat.

Development of a Prismoid

The prismoid, as shown in figure 10.6, is selected for the first example. Carefully note the general shape of the solid, especially the shape of the sides. Draw in a base line and lay off points 1, 2, 3, 4, and 1, as shown in figure 10.7. The distances between these points are obtained from the plan. The distance between points 1 and 2 in the sketch is equal to the distance between points 1 and 2 in the plan; the distance between points 2 and 3 in the sketch is equal to the distance between points 2 and 3 in the plan, and so on all the way around to the starting point 1. Erect perpendiculars at the points thus obtained, and project points *A*, *B*, *C*, *D*, and *A*, over from the elevation by the dotted lines parallel to the base line.

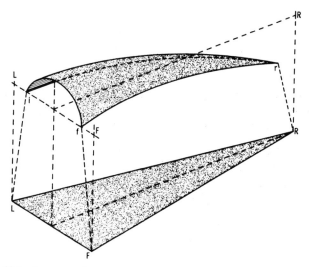

Fig. 10.5. An approximate pattern as cut for a section of a warped surface, showing shape of pattern before and after warping.

The intersections of these dotted lines with the perpendiculars will give the heights of the perpendiculars corresponding to the heights of the edges of the prismoid.

Example *Develop a pattern for the cylindrical surface of the cut cylinder shown in figure 10.8. In figure 10.9, take any number of*

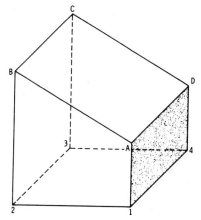

Fig. 10.6. A prismold in cabinet projection.

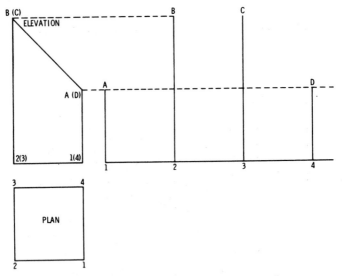

Fig. 10.7. Development of a pattern by parallel lines.

points on the circle, such as 1, 2, 3, 4, 5, 6, 7, and 8, and project up in elevation, obtaining points A(A), B(H), C(G), D(F), and E. Here some of the points coincide; that is, some lie directly back of the others. Thus, point 2 or its projection (B) lies back of 8 or its projection H, the parentheses indicating the fact. Draw the base line for the stretch-out, and lay off on this line points 1, 2, 3, 4, 5, 6, 7, 8 and 1, spaced equal to the lengths of the arcs between points 1, 2; 2, 3; etc., in the plan.

In the stretch-out, erect perpendiculars through the points 1, 2, 3, etc., and project over from elevation, points A, B(H), C(G), D(F), and E, making points A, B, C, D, E, F, G, H, and A. Draw a curve through these points, which completes the stretch-out. Note carefully in projecting over points from elevation their location on the stretch-out. Thus, point (B) is projected to the perpendicular through 2, and point H to the perpendicular through 8; this is evident from the plan. Where a large number of points are taken, the length of the spacing of the points in the stretch-out may be obtained by setting the dividers by the chord method, as in figure 10.10. It should be understood that this is only an approximation, and the stretch-out will never be the full length, no matter how many points are taken.

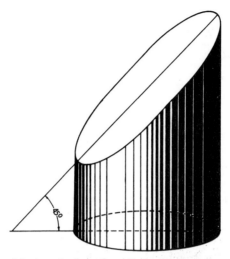

Fig. 10.8. A cabinet projection of a cylinder cut by a plane inclined 45° to the base.

For precision, especially where only a few points are taken (as in figure 10.9), the dividers should be set by calculation, which is called *line setting* (figure 10.11). A comparison of the two methods is shown in figure 10.12. To illustrate the method of line setting, suppose the diameter of the cylinder is 6″, and the 8 points are taken as shown in figure 10.11. Then

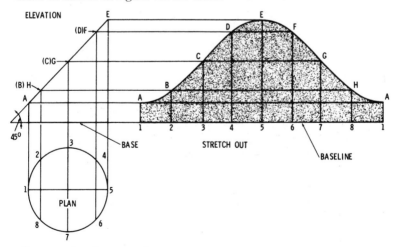

Fig. 10.9. Development of a pattern by parallel lines for the cylinder cut by 45° plane.

135

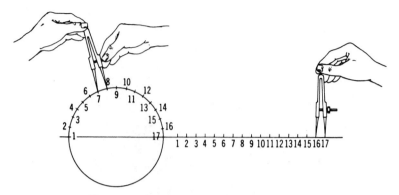

Fig. 10.10. Chord method of spacing.

$$\text{circumference} = 6 \times 3.1416 = 18.85''$$

The length of the stretch-out is equal to 18.85″, and the distance between points equal to $18.85 \div 8 = 2.36''$. Draw a line and accurately measure off a distance *LF* equal to 18.85″. Now, set the dividers to

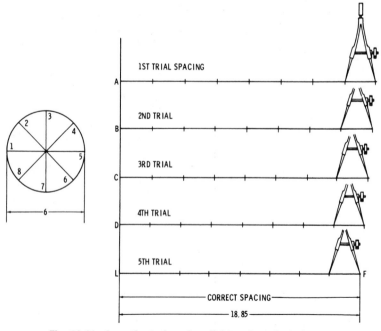

Fig. 10.11. A method of setting dividers by calculations.

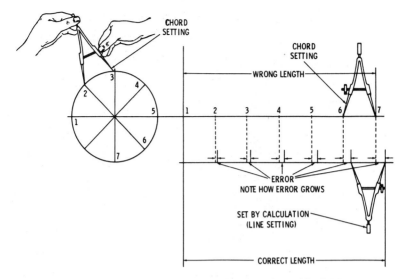

Fig. 10.12. Comparison of chord and line setting with dividers.

2.36″ and make a trial spacing. The result obtained will probably be as on line A. Note the magnitude of the error. Make additional trials, as indicated on lines B, C, D, etc., until the true setting is obtained, as on line LF.

Figure 10.13 illustrates a parallel line development for a small scoop. This parallel line development can be drawn to different dimensions utilizing the steps illustrated in figure 10.9. This process can be used for a number of different sheet metal applications in heating and air conditioning work.

RADIAL LINES

The second class of elementary surfaces are those whose elements are not parallel but radial from a common point.

What shape of objects have surfaces with radial lines?
    Answer: Pyramids, cones, etc.

Name two subdivisions of radial line developments.
    Answer: (1) Surfaces having three or more consecutive elements in the same plane, and (2) a surface not having three consecutive elements in the same plane.

137

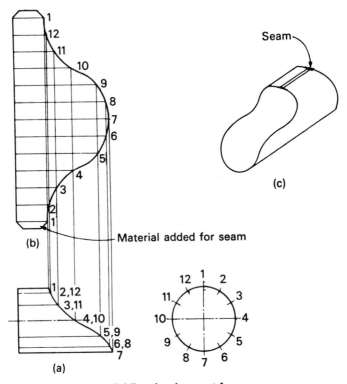

Fig. 10.13. Parallel line development for a scoop.

Give an example of objects belonging to the first group.
   Answer: Pyramids.

Give an example of objects belonging to the second group.
   Answer: Cones.

Radial-Line Developments

The following examples are given to illustrate the method of development by radial lines.

Problem 1—Develop by radial lines the pyramid shown in figure 10.14.

What should be drawn first?
   Answer: An elevation and plan, as shown in figure 10.15.

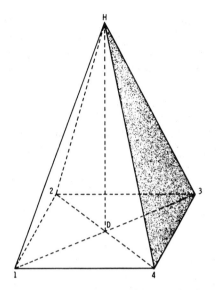

Fig. 10.14. A pyramid in cabinet projection.

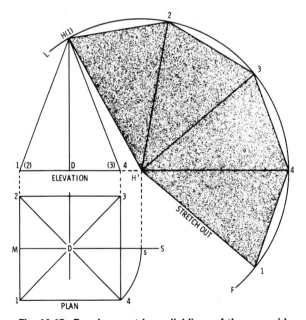

Fig. 10.15. Development by radial lines of the pyramid.

139

What should be noted on examination of the pyramid?

   Answer: It will be seen that the edges $H1$, $H2$, $H3$, and $H4$, are of equal length, and since these edges (which are elements of the surface) converge to a common point, their extremities in the development will lie in an arc of which the common point or apex is the center.

What should be noted about the elevation?

   Answer: The elevation does not show the true length of the elements at the four corners.

What must be done to find the length of the elements?

   Answer: Revolve $H4$ into the plane of the elevation.

How is this done?

   Answer: Describe an arc through point 4 in plan with $D4$ as the radius and where this cuts axis MS at s, project up to H' and draw HH', which gives the true length of the elements.

How is the surface developed?

   Answer: Take H' as center and, with $H'H$ as a radius, describe arc LF. Set the dividers to the distance of one side obtained from plan, and space off points 1, 2, 3, 4, and 1 on LF. Connect these points with H', and draw lines connecting points 1, 2; 2, 3; 3, 4; and 4, 1. The figure thus obtained is the developed surface of the pyramid.

Problem 2—Develop by radial lines, the conical hood of a smoke stack, as shown in figure 10.16.

What is the shape of the plan of the hood?

   Answer: A circle.

How is the elevation obtained?

   Answer: The elevation will be a triangle whose base equals the diameter of the plan and whose altitude equals the height of the cone, that is, from base to apex.

How is the surface of the cone developed?

   Answer: Divide the base of the circle into any number of equal parts, say 36. Using point D as the center on the pattern, and with the radius DE equal to AC, describe an arc starting at point E. In succession, divide the circle base into 36 equal parts in the same manner as accomplished on the plan. The space at points EF is the portion to be removed to develop the cone.

140

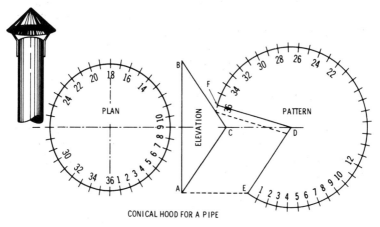

CONICAL HOOD FOR A PIPE

Fig. 10.16. A conical hood on smoke stack, and the development of its surface.

What are the boundaries of the required development?

Answer: The two radial lines from point *D* in figure 10.16, and the arc that is equal in length to the circumference of the base circle of the cone.

Problem 3—Develop by radial lines the oblique cone as shown in figure 10.17.

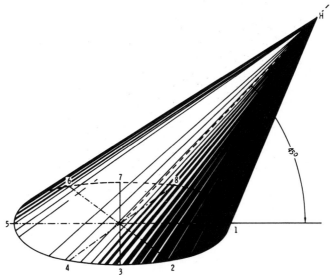

Fig. 10.17. An oblique cone in cabinet projection.

141

What is drawn first?

> *Answer:* The elevation and plan, as shown in figure 10.18.

What element appearing in its true length is taken as the beginning of the development?

> *Answer: H*1, shown in figure 10.18.

Do the other elements appear in true length?

> *Answer:* With the exception of enement *H*5, the others do not appear in true length.

Describe the various steps in developing the surface.

> *Answer:* Revolve *H*′2 to *MS* in plan, and project up to 2′ in elevation, giving *H*2′ as the true length of *H*2.
>
> Similarly, obtain *H*3′, and *H*4′ as true lengths of *H*3 and

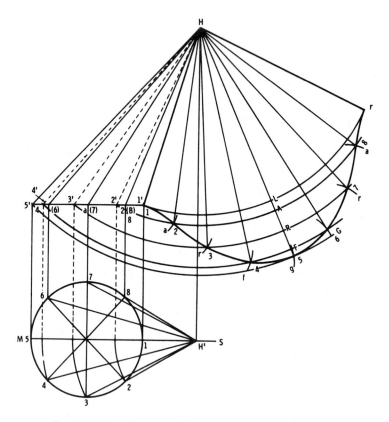

Fig. 10.18. Development of the oblique cone by radial lines.

H4. With *H* as the center, and *H*1 as the radius, describe arc *L*. With radius *H*2′, describe arc *A*; with radius *H*3′, describe arc *R*; with radius *H*4′, describe arc *F*; with radius *H*5′, describe arc *G*. Set the compass to the common distance between the elements in the plan, as the distance between points 1 and 2. With dividers set to this distance, and with 1 (in stretch out) as the center, describe arc *a* cutting *A* at 2. With 2 as the center, describe arc *r*, cutting *R* at 3. With 3 as the center, describe arc *f* cutting *F* at 4. With 4 as the center, describe arc *g* cutting *G* at 5. This gives points for half of the pattern; the other half is similar. Points 6, 7, and 8 are obtained by the intersection of arcs *f′r′a′* with *F′RA*. Join all of the points by a curve, and draw 1′*H*, thus completing the pattern.

Problem 4—Develop by radial lines the conical eave trough outlet shown in figure 10.19. It does not matter what curve is given

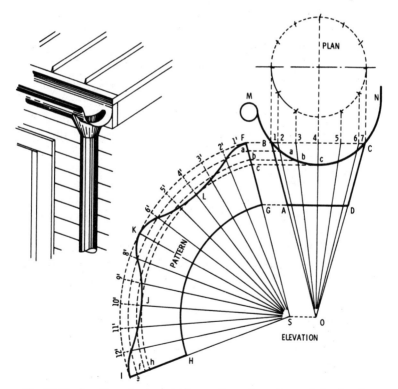

Fig. 10.19. A pattern for a conical eave through outlet, and the development of its surface.

to the gutter—oval or circular. The method explained in this case is suitable for gutters of all curvatures. In this instance, the cross section of the gutter is represented by *MBCN* in elevation. The conical outlet, which is to form a connecting piece between the gutter and the pipe, is a truncated cone whose wider end is shaped so as to conform with the surface of the gutter. The cone in which the connection piece is a part is represented as *BOC*. The dotted line *BC* represents the diameter of the base of this cone; the circumference is shown in the plan.

Describe the method of development.

Answer: First, develop the truncated cone *ABCD*. Obtain the figure bounded by the radial lines *HI* and *GF* (each equal to *AB*) and by the arc *IKF* and *HG*. The larger arc is described with a radius equal to *OB*, while the smaller arc has a radius equal to *OA*.

What is the length of the larger arc?

Answer: It is equal to the sum of all the divisions on the circumference of the circle shown in the plan.

How are these divisions marked?

Answer: They are marked on the development by the points 1', 2', 3', etc.

The points 1, 2, 3, etc., on the base line *BC*, are connected with the vertex *O* by straight lines. The surface of the cone appears with a series of elements or lines converging in the vertex. On the development of the cone, these elements appear as the radiating lines *S1'*, *S2'*, *S3'*, etc. Where the elements of the cone are intercepted by the cross section of the gutter, there are the points *B*, *a*, *b*, and *c*, which are projected to the line *FG* by projectors parallel to *BF*, thus marking off on the line *GF* the points *F*, *a'*, *b'*, and *c'*. From these points describe the arcs *a'e*, *b'f*, and *c'h*. These arcs, in their intersections with the lines *S1'*, *S2'*, *S3'*, etc., define the curved outward boundary *FLKJI* of the required pattern.

What modification should be made in making an actual pattern for the trough outlet?

Answer: The pattern should be enlarged along the curved boundary by the addition of stock for laps or locks.

Problem 5—Describe the procedure used in developing a radial line development for a funnel (see figure 10.20).

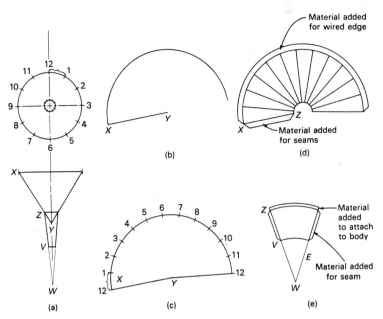

Fig. 10.20. A radial line development for a funnel.

Answer: The procedure used in drawing a radial line development for a funnel is to make two views of the funnel showing its shape and size. After this has been done, draw the layout with *XY* as the radius. Transfer the equal spaces representing the circumference in the top view of the layout. Next, draw the arc that forms the lower edge of the funnel body. Use the distance *XZ*. Draw seams onto the sides of the body of the layout. Next, draw the allowance for the wired edge. This allowance is calculated at two and one-half times the diameter of the wire that is used in the wired edge. After this has been done, draw the layout for the spout following the same procedure that was used to draw the body. After the layout has been completed, check its accuracy by cutting it out of the paper and folding it to shape.

TRIANGULATION

There are some forms of elementary surfaces so shaped that, although straight lines can be drawn on them, such lines when drawn would not be parallel nor inclined toward each other with any degree of accuracy.

What kind elements ares are not contained on such surfaces?

Answer: They contain neither parallel nor radial elements.

What can be drawn on such surfaces?

Answer: Two or more elements can be drawn in certain directions, forming angles.

On such irregular surfaces, it may happen that no two of the angles thus drawn on the solid, or represented in the projection drawing, will lie in the same plane or be equal to each other. Since it is possible to project these angles, they can be reproduced on the flat surface of the drawing paper in their correct size. If this can be done, it may be reasonably assumed that the surfaces thus represented will be the same as the corresponding surfaces belonging to the class and developed by the method of triangulation. To the student who thoroughly understands the principles of projection, the theory of triangulation should not present any serious obstacles.

What are the steps in developing a surface by triangulation?

Answer: First, the surface is divided up into a number of elementary triangles (called triangulation). Next, the true lengths of the sides of the triangles are found, and the triangles reproduced in the pattern.

Example *Develop by triangulation the irregular shaped object shown in figure 10.21.*

After drawing an elevation and plan as shown in figure 10.22, what is the next step?

Answer: Select any number of points on the bottom edge and the same number of similarly located points on the top edge. For simplicity only, 8 points are taken on each edge (though in practice a greater number are taken).

How are the points similarly located?

Answer: If the points are taken, for instance, at the intersection of similar axes with the edges, they will be similarly located. These points are 1, 3, 5, and 7 for the bottom, and A, C, E, and G for the top.

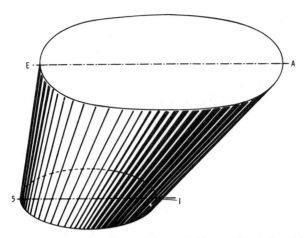

Fig. 10.21. An irregular object with parallel bases shown in cabinet projection.

What is the next step?

 Answer: Determine the true lengths of the elements by constructing, for each element, a right triangle whose base is equal to its projection on the base or length in the plan, and its altitude equal to the vertical height of the element in elevation. The hypotenuse of such a triangle will then equal the true length of the element.

How are the triangles constructed?

 Answer: Continue over to the right, top, and base in elevation. The distance between these lines will equal the common altitude of the triangles. Beginning with element $A1$, its true length appears in elevation. It is not necessary to construct a triangle to find its true length. For the next element $B1$, set the dividers to distance $B1$ in the plan and mark this distance on the base line of the triangle layout as $b1$. Draw perpendicular bB and join $B1$, thus completing the triangle. Its hypotenuse $B1$ is then the true length of the element $B1$, which appears foreshortened in both plan and elevation. In a similar manner, the true lengths of all the other elements are found. Next, lay out the pattern, using the true lengths of the elements just found.

Example *Develop by triangulation the irregularly shaped object shown in figure 10.23. This is virtually the same problem as worked*

147

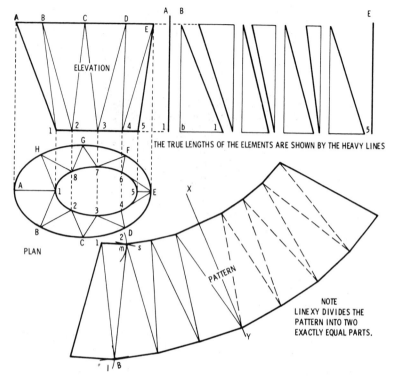

Fig. 10.22. Development by triangulation of the irregular shaped object shown in figure 10.21.

out in figure 10.10, except that the top is inclined to the base. The same method is used and the drawings of each problem are similarly lettered. The only difference in laying out the lines is in the triangle layout. Here, as will be seen in figure 10.24, the triangles have different altitudes, depending on the location of the points on the top edge. Elements A1 and E5 appear in their true length; no triangles are necessary for these.

How is the development or pattern laid out?

Answer: With A as the center, and the radius equal to the chord distance AB in the plan, describe arc 1. With 1 as the center, and with the radius equal to the true length of element B1 (as found in the triangle layout), describe arc f, intersecting arc 1 at B. Join A and 1 to B, thus completing the first triangle A1B.

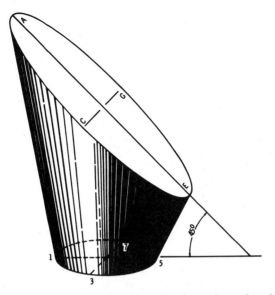

Fig. 10.23. An irregular object with inclined top shown in cabinet projection.

For the second triangle, take point 1 of the pattern as the center, and, with a radius equal to the chord distance 12 in the plan, describe arc m, and with B as the center and with a radius equal to the true length of element $B2$ (obtained from the triangle layout), describe arc s, intersecting m at 2. Join 1 and B to 2, thus completing the second triangle 1B2. Continue in the same manner until the pattern is completed.

WARPED SURFACES

By definition, a warped surface is one in which a straight edge may be placed in contact only at a point. It is not possible to develop a warped surface.

Why is it not possible to develop a warped surface?
    Answer: Because the surface does not contain elements.

After developing a pattern approximating a warped surface, what should be done to make the pattern coincide with the surface?
    Answer: It is necessary to raise the surface of the metal pattern by hammering to shape, so that when the pattern is in position it will fit.

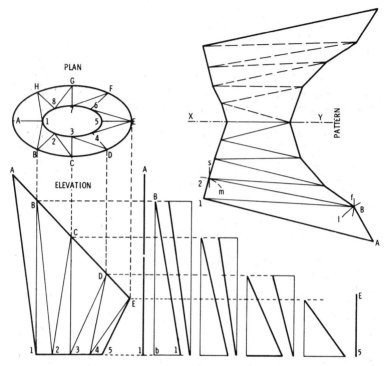

Fig. 10.24. Development by triangulation of the irregular shaped object shown in figure 10.23.

Give a typical example of a warped surface.

    Answer: A sphere.

Pattern Development of a Hemisphere

How is the warped surface divided?

    Answer: It is divided into as many sections as desired. The pattern of each section will approach the shape of the warped surface when placed in position; that is, a less amount of hammering will be necessary to raise the surface of the pattern so it will coincide with the warped surface.

In dividing the hemisphere into sections, what do the sections consist of?

    Answer: They may either be *zones*, or *segments*.

Zone Method—In figure 10.25, draw the elevation of a hemisphere and divide it into zones A, B, and C.

What are the zones A and B?

Answer: Frustums of cones.

What is zone C?

Answer: A cone.

What is the first step in development of frustum A?

Answer: Continue the slant surface of frustum A until it intersects the axis at H', giving a radius center for frustum A. With H' as the center, describe the two arcs M and S.

What is the object of the end view?

Answer: It gives the boundaries of the zones.

How is the length of the development obtained?

Answer: In the end view rectify arc 12, and with this rectified arc length space off on arc S the points 1, 2, 3, 4, and 1.

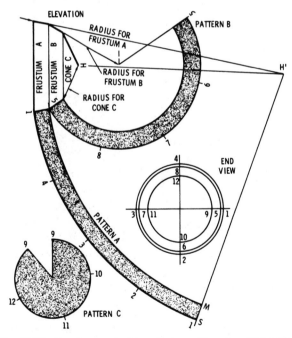

Fig. 10.25. Development of hemisphere. Case 1: zone method.

How is the pattern or development completed?

Answer: By connecting *M* and *S*, that is, by extending the radial line *H'M* to *S*.

How are frustum *B* and cone *C* developed?

Answer: In a similar manner, as indicated in figure 10.25.

What is the appearance of the three patterns when assembled?

Answer: They appear as shown in figure 10.25.

What must be done to bring the assembly to the shape of a hemisphere?

Answer: The metal of the patterns must be raised to the warped shape of the hemisphere by hammering.

Segment Method—Figure 10.27 shows one fourth of a hemisphere divided into two segments. The shaded surface 1*H*2 is one of these sections. It is only necessary to develop a pattern for one section since all the others are of the same shape.

How does figure 10.28 show the hemisphere?

Answer: Partly in cabinet projection and partly in elevation.

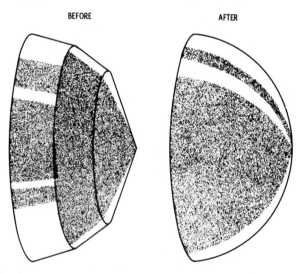

BEFORE   AFTER

Fig. 10.26. Appearance of pattern for hemisphere before and after hammering or raising to the warped shape of the hemisphere. Case 1: zone method.

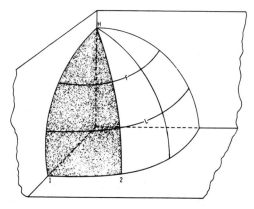

Fig. 10.27. Quadrant of a hemisphere showing surface divided into segments.

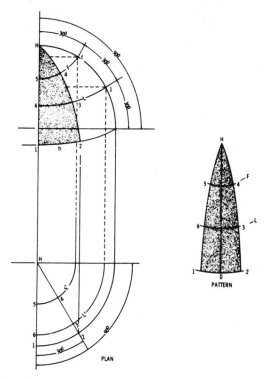

Fig. 10.28. Development of patterns by approximation for a hemisphere of warped surface. Case 2: segment method.

153

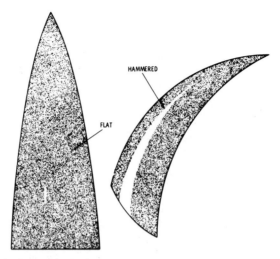

Fig. 10.29. Appearance of pattern for hemisphere before and after hammering. Case 2: segment method.

What should be drawn first?

Answer: The zone circles *F* and *L*, spaced as shown in figure 10.28.

What is drawn in the plan in figure 10.28?

Answer: Draw projection 1*h*2 of the segment for which a pattern is to be developed. Project also into plan points 1 and *f*, and describe zone *L'* and *F'*, which is the projection in plan of *L* and *F* in elevation.

Describe in detail how to develop the pattern.

Answer: Draw an axis *HD*. Rectify arc *Hh*, and mark off the rectified length as *Hh* on the pattern. With *H* as the center, describe an arc through *h*, and also arcs *L* and *F*, dividing *Hh* into three equal parts. On the pattern lay off 1 and 2 equal to 1 and 2 in the plan; 6 and 3 equal to 6 and 3 in the plan; 5 and 4 equal to 5 and 4 in the plan. Through the points thus obtained draw curves connecting *H* to 1 and 2, thus completing the pattern.

What does the pattern look like before and after hammering?

Answer: As shown in figure 10.29. Note that the pattern is flat before hammering.

CHAPTER 11

Detail Drawings

The difference between *assembly drawings* and *detail drawings* was pointed out in a preceding chapter. The student should now learn drafting procedure with respect to detail drawings—how many views are necessary, what views, sections, sectional views, dimensioning, information, data, etc. In the several examples given in this chapter, the object will be shown in a pictorial drawing, such as cabinet or isometric projection. The problem is to make the necessary detail drawings from which a mechanic could make or machine the required object.

Problem 1—Make a detailed drawing of the open-top rectangular object shown pictorially in figure 11.1.

What are the minimum number of views required to complete the drawing?
 Answer: Two.

What two views would completely show the object?
 Answer: A cross section and a longitudinal section, as shown in figure 11.2.

Could any other view be added to make the print more easily readable?
 Answer: The addition of a top plan, while not necessary, would help to give a mental picture of the shape of the object.

How is width *AB* of the inclined side obtained?
 Answer: Either by direct measurement on the drawing or by calculation, as shown in figure 11.3.

Which is more accurate?
 Answer: By calculation.

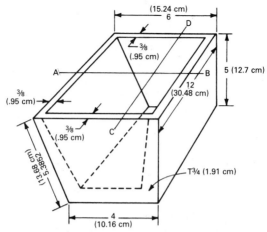

Fig. 11.1. Pictorial view of a rectangular open-top object.

Why?

Answer: Because it is physically impossible to measure anything exactly.

Problem 2—Make the necessary detail drawings of a water faucet (shown in fig. 11.4) and indicate the dimensioning without putting on actual dimensions (see fig. 11.5).

How many detail drawings are required?

Answer: Fourteen, counting all views.

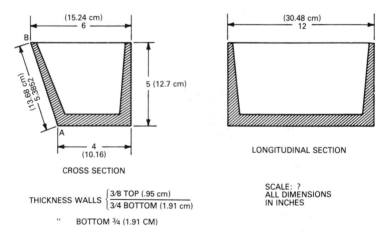

CROSS SECTION

LONGITUDINAL SECTION

THICKNESS WALLS $\begin{cases} 3/8 \text{ TOP (.95 cm)} \\ 3/4 \text{ BOTTOM (1.91 cm)} \end{cases}$

"   BOTTOM ¾ (1.91 CM)

SCALE: ?
ALL DIMENSIONS
IN INCHES

Fig. 11.2. Cross section and longitudinal action drawings of the object shown in figure 11-1.

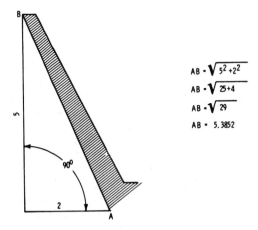

$$AB = \sqrt{5^2 + 2^2}$$
$$AB = \sqrt{25 + 4}$$
$$AB = \sqrt{29}$$
$$AB = 5.3852$$

Fig. 11.3. Mathematical method of obtaining width of inclined side of the object.

Why so many?

Answer: As indicated in the assembly drawing (fig. 11.4), there are eight parts to the valve, and some of them require more than one view. Two elevations are required for the faucet body, three views for the shank, two for the valve, one each for the valve rod and spindle, and one for the handle.

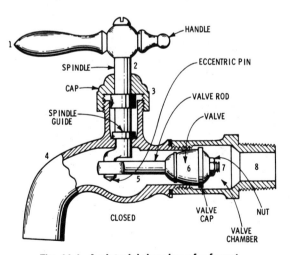

Fig. 11.4. A pictorial drawing of a faucet.

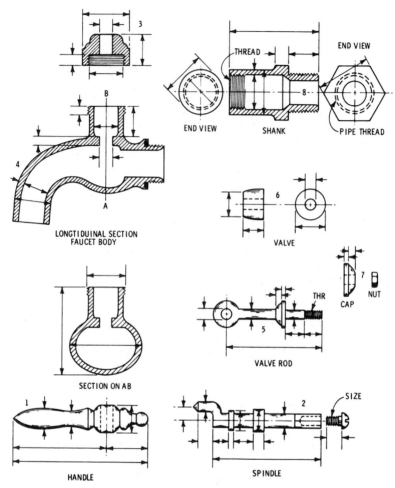

Fig. 11.5. Detailing drawings of the water faucet as shown in figure 11.4.

What other data should be on the drawings?

Answer: The material used in each part, and also the pitch of the various threads.

Why is the scale omitted?

Answer: Because of the reduction in size of the original drawings for this book, the scales would have no meaning. A scale gives no indication of actual sizes on the drawing when reduced or enlarged from the size of the original drawing.

158

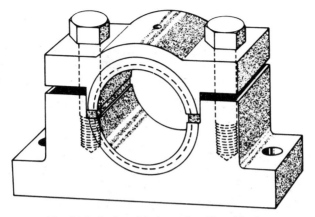

Fig. 11.6. A pictorial view of a pillow block.

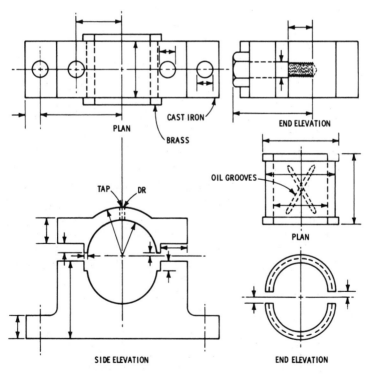

Fig. 11.7. Detail drawings of the pillow block shown in figure 11.6.

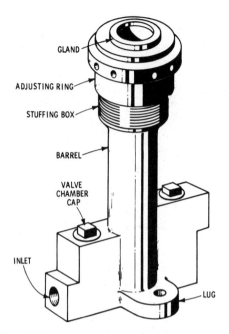

GLAND

ADJUSTING RING

STUFFING BOX

BARREL

VALVE
CHAMBER
CAP

INLET

LUG

Fig. 11.8. Pictorial drawings of a small "direct connected" boiler feed pump.

Problem 3—Make a detail drawing of the pillow block shown pictorially in figure 11.6.

What is a pillow block?

Answer: A box or frame enclosing and supporting a brass journal or bearing in which a shaft revolves. Generally, it consists of two parts—the box or block holding half of the brass and the cup holding the other half, the two halves being adjustable for wear.

How many detail drawings would ordinarily be made to completely show the pillow block in figure 11.6?

Answer: Five views, as shown in figure 11.7.

How should these drawings be dimensioned?

Answer: As indicated in the various views (see fig. 11.7). Identical dimensions for symmetrical parts referred to as main axis should be omitted as they are unnecessary and clutter up the drawings.

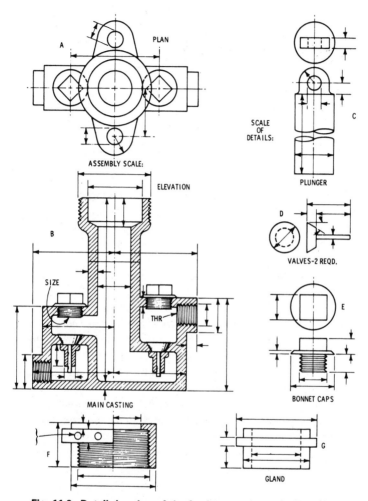

Fig. 11.9. Detail drawing of the feed pump shown in figure 11.8.

Problem 4—Make detail drawings for the small pump shown pictorially in figure 11.8.

How many detail drawings should be made to completely show the boiler feed pump?

Answer: Seven views, as shown in figure 11.9.

How should the detail drawings be dimensioned?

Answer: As indicated in the various views shown in figure 11.9.

CHAPTER 12

Sketching

Sketching is a valuable tool to the mechanic, the engineer, and the drafter. It is a freehand form of mechanical drawing. The words *sketching* and *freehand* may suggest carelessness or incompleteness, but this is not the case. The difference between a drawing and a sketch is that the sketch is not as refined as the drawing. It is a quick representation of an idea that the sketcher is attempting to put on paper.

Often the mechanic will find himself making sketches to supplement drawn plans in order to clarify some part or section. Most drawings are started from sketches (the original ideas were sketched out) and the dimensions added, and then completed drawings are made later from the sketches. A minimum of materials is needed for sketching. Plain or squared (graph) paper, a medium pencil, eraser, clip board, and a small pocket ruler are valuable items. A ruler and calipers will be needed to take dimensions from the item that is being sketched. At the start, plain paper should be used in order to avoid a later dependency upon lines in making good legible sketches.

To what does the term *sketching* especially apply?
   Answer: To pictorial outlining.

What term for sketching would probably be more specific here?
   Answer: Freehand drawing.

What are its applications?
   Answer: It is used to make rough orthographic drawings without the aid of drawing instruments, such as a T square or triangles. The ability to make a clear sketch quickly is a valuable asset to the drafter and mechanic as well. Either may be called upon to sketch some details of a machine that is to be altered,

or upon which some improvement is to be made. Clearly such sketches are of no use to the designer unless they contain all the dimensions and data necessary for the machinist or tool-maker to redesign the part.

What hardness of pencil lead should be used?

Answer: It depends upon the character of the drawing and the precision required. Some drafters recommend a 2H pencil, but this is pretty soft for accurate work. A 4H is better, and a 6H is used sometimes.

What is the objection to hard pencils?

Answer: The harder the pencil lead the more difficult it is to see the lines.

HOW TO SHARPEN PENCILS

There are two methods of sharpening a drawing pencil, classed with respect to the shape given the lead: (1) flat or chisel, and (2) conical.

What is done preliminary to sharpening the lead?

Answer: The wood is cut down to a cone shape, being careful not to cut the lead.

Describe the shapes of lead generally used.

Answer: Figure 12.1 shows these shapes.

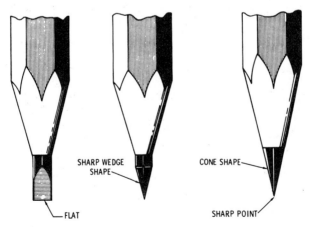

SHARP WEDGE SHAPE

CONE SHAPE

FLAT

SHARP POINT

Fig. 12.1. Side and end views of a pencil with the lead sharpened to a chisel, wedge, or conical shape.

How is a pencil sharpened to the flat or chisel shape?

Answer: Slide the lead (at a very acute angle) along the pad, forming it to the chisel shape shown in figure 12.2.

How is a pencil sharpened to a conical or round point?

Answer: Slide the lead over the pad in one direction, turning it on its axis during the movement along the pad, as shown in figure 12.2. This causes even wearing of the lead.

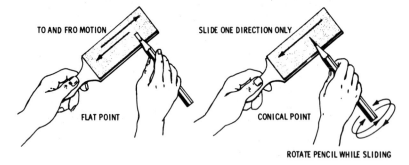

Fig. 12.2. A method of sharpening pencil for flat or conical point.

How are newly sharpened pencil points cleaned?

Answer: One of the quickest ways to clean newly sharpened pencil points is to use a disc desk top lead pointer cleaner (see fig. 12.3). The pencil point is inserted into the cleaning area.

Fig. 12.3. Desk top lead pointer cleaner.

What might be said with respect to a pencil with a dull point?

Answer: Do not expect to do good work with a pencil having a dull point.

What movement should be given to a pencil with a conical point in drawing lines?

164

Answer: After drawing a line, slightly rotate the pencil before drawing another line. This helps to keep the pencil sharp.

METHOD OF HOLDING A PENCIL

Much depends upon holding the pencil correctly. The proper way is shown in figure 12.4.

The pencil should be held firmly between the thumb and first finger of the right or left hand. Press the second finger against the pencil at the opposite side to the thumb pressure, so that the pencil is firmly held by the contact of the thumb and two fingers, the third and fourth fingers just coming into easy reach of the paper surface— the wrist or ball of the hand resting lightly on the surface of the work, the arm resting on the desk or drawing board for steadiness.

How is the motion of the pencil produced for vertical stroke?

Answer: It is produced from the movement of the fingers and thumb.

For horizontal strokes?

Answer: By fingers and thumb combined with a wrist or elbow motion.

How are oblique lines produced?

Answer: By a free movement of the finger joints.

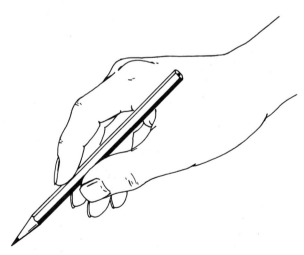

Fig. 12.4. Approved position of holding pencil for freehand drawing.

What precaution should be taken in holding the pencil?

Answer: Do not hold the pencil too close to the end.

Why?

Answer: By letting it project sufficiently beyond the fingers, the movement given by the fingers is multiplied and this helps in drawing straight lines. The pencil should be held perpendicular to the paper and inclined to the right. Draw horizontal lines from left to right; vertical lines from bottom upward.

HOW TO DRAW FREEHAND

It is essential for the drafter to be able to draw straight lines and round circles in order to attain proficiency in making clear and legible sketches. He must be able to draw lines in various positions, for example,

1. Horizontal
2. Vertical
3. Oblique to right
4. Oblique to left

Horizontal Lines

Hold the pencil as in figure 12.5, keeping the elbow near the side. Produce the line by one light steady stroke, the movement being obtained by motion of the wrist.

How is the required width of the line obtained?

Answer: By going over it one or more times. This method induces lightness of touch and freedom of movement. Evidently it has the further advantage that if the initial line is not right it is easier to erase than a heavy line. Follow this method until a heavy line can be drawn without the necessity of erasing.

What kind of movement should be given to the pencil with the single-stroke method?

Answer: The movement should be even throughout, not made with a jerky movement.

Describe the progressive method of drawing lines.

Answer: In this method, the line is drawn by several short

166

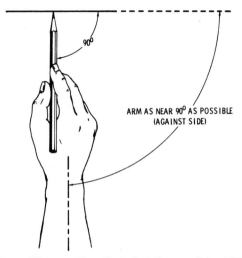

90°

ARM AS NEAR 90° AS POSSIBLE
(AGAINST SIDE)

Fig. 12.5. Horizontal line position. Note that the pencil is at 90° to the line to be drawn.

strokes, each stroke producing part of the line. The student should practice both methods just described, adopting the one that presents the least difficulty.

Right Oblique Lines

Hold the pencil as shown in figure 12.6, with the elbow a little from the side. Draw the line with one light stroke by a movement of the fingers and thumb, replacing the strokes until the line is as heavy as desired.

What should be avoided in drawing oblique lines, and why?

Answer: Do not draw an oblique line using the wrist or elbow as a hinge with the fingers rigid. The tendency of such movements is to produce arcs.

What does wrist movement produce?

Answer: An arc of short radius.

How is an arc of long radius produced?

Answer: By using the elbow as a hinge for the movement.

How may the proper movement be detected?

Answer: By watching the action of the thumb. If it bends as the line is being drawn, the correct movement is obtained.

167

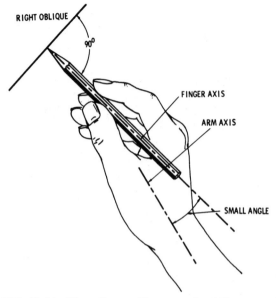

Fig. 12.6. Right oblique line position, arm should be at small angle.

Left Oblique Lines

Hold the pencil as in figure 12.7, with the elbow removed far from the side. The correct distance is obtained when the arm is at an angle of 90° to the line to be drawn.

How are left oblique lines drawn?

> *Answer:* With a movement of the fingers and thumb.

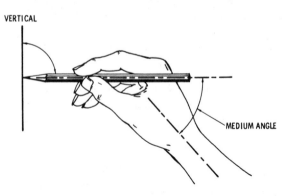

Fig. 12.7. Vertical line position, arm at medium length.

Vertical Lines

To draw a vertical line requires more care than the lines just described. Hold pencil as in figure 12.7, with the elbow moved well out from the side, and draw the line by a movement of the fingers and thumb.

What is the usual tendency in drawing freehand vertical lines?

Answer: The tendency in drawing vertical lines is to slightly incline to the horizontal instead of the perpendicular.

How is this tendency overcome?

Answer: Place the paper in such a position that when the line is drawn it will be exactly in front of the eyes.

It will be noticed that each change in direction of the line to be drawn has been accompanied by a corresponding change in the position of the elbow and wrist. Note the following with respect to position of the elbow:

1. Horizontal line, elbow near side
2. Right oblique line, elbow a little removed
3. Vertical line, elbow more removed
4. Left oblique line, elbow most removed

In general, the position of the pencil should be at right angles (90°) to the line to be drawn, as shown in figure 12.8. The pencil should be held perpendicular to the paper and inclined to the right. Draw vertical lines from the bottom upward.

LINE-PRODUCING MOVEMENTS

With respect to the movements by which lines are drawn, they may be classed as (1) finger and thumb lines or (2) wrist lines. In addition, there are lines that may be drawn by either finger and thumb, or wrist movement.

How are lines 45° to 90° from the horizontal drawn?

Answer: By finger and thumb movement, as shown in figure 12.9A. Lines 45° from the horizontal could be drawn as shown in figure 12.9B.

How are horizontal lines drawn?

Answer: By wrist movement, as shown in Fig. 12.10.

169

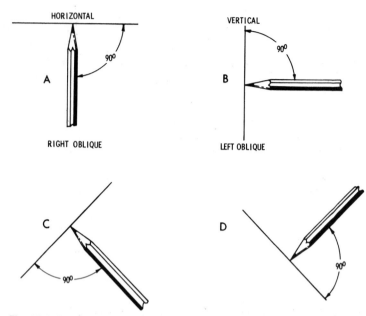

Fig. 12.8. Position of the pencil is always 90° to the line to be drawn, showing (A) horizontal, (B) vertical, (C) right oblique, and (D) left oblique.

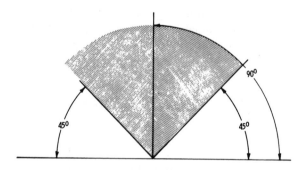

Fig. 12.9A. Finger and thumb lines.

DIRECTION OF PENCIL MOVEMENT

In drawing lines, the direction in which the pencil is moved depends upon the inclination or final direction of the line to be drawn: (1) horizontal lines from left to right and (2) vertical lines upward.

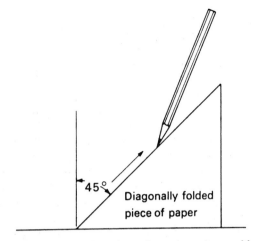

Fig. 12.9B. A diagonally folded piece of paper can be used in sketching a 45° angle.

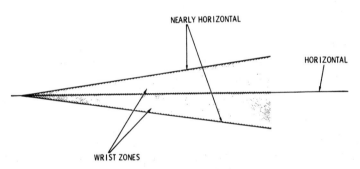

Fig. 12.10. Wrist lines.

ARCS AND CIRCLES

The correct position of the hand in describing an arc is the same as that required for drawing a straight line joining its extremities, as shown in figure 12.11. Note that the pencil is held inside or outside of the arc, according to the position of the arc. There are several methods of describing a circle:

1. Guide marks on axes

2. Inscribing in a square

3. Preliminary pencil movement

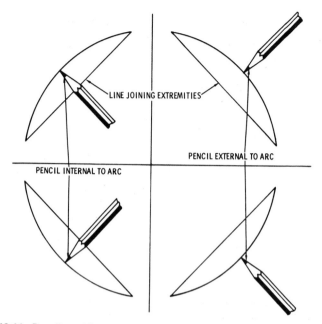

LINE JOINING EXTREMITIES

PENCIL EXTERNAL TO ARC

PENCIL INTERNAL TO ARC

Fig. 12.11. Pencil position used in describing arcs in four quadrants.

Describe the first method.

Answer: In figure 12.12, draw the rectangular axes XX and YY, and space them off by eye in equal distances $o1$, $o2$, etc., each equal to the radius of the circle to be described. Then describe the circle through these points.

What should be done for greater accuracy?

Answer: Draw the diagonal axis indicated by dotted lines, spacing them off at oa, ob, etc. (which is the same length as the other radius), thus obtaining more guide points.

Is there an easier method of describing the circle after the foregoing preliminary?

Answer: Turn the paper around for each quadrant instead of changing position of the hand. It is not always convenient to shift the paper, so the student should acquire sufficient technique on circles so as not to resort to shifting the paper.

Describe the square method of describing circles.

Answer: Draw a square $ABCD$ (figure 12.12) symmetrical

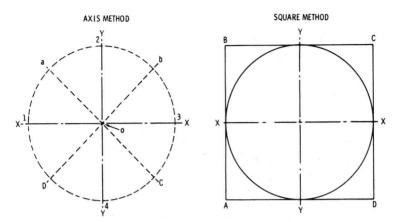

AXIS METHOD

SQUARE METHOD

Fig. 12.12. Square method in describing a circle.

with the rectangular axes *XX* and *YY*. This is made up of four little squares, each of which will serve as a guide in describing the circle progressively with 90° arcs, the completed circle being tangent to the four sides of the large square.

Describe the preliminary pencil movement method of describing circles.

Answer: As shown in figure 12.13, it consists of rotating the pencil several times before contact with the paper, the radius of rotation being that of the desired circle.

In the case of a small circle, this can be done by finger movement, but for a large circle, the hand should be rigid and the movement obtained from the entire arm with the elbow as a hinge. In the preliminary rotation, make as many turns as necessary until the pencil point describes the nearest approach to a circle of the desired diameter, then let the pencil point make contact with the paper.

Describe the methods of preliminary movement practice.

Answer: Practice with clockwise movement, and then with counterclockwise movement, adopting the one that is easier. The physical characteristics of drafters differ, and rigid rules cannot be laid down for all operations.

Describe the method using scrap paper to mark off the radius points.

Answer: Scrap paper can be used to mark off equal dis-

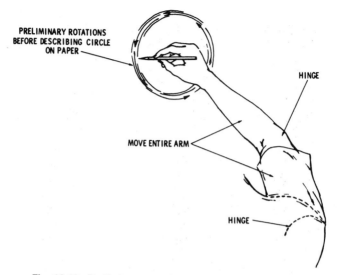

PRELIMINARY ROTATIONS
BEFORE DESCRIBING CIRCLE
ON PAPER

HINGE

MOVE ENTIRE ARM

HINGE

Fig. 12.13. Preliminary pencil movement in describing a circle.

tances when sketching circles and arcs (see fig. 12.14). This aid is useful when marking off radius points. Connect the radius points with light construction lines and then darken in the circle.

Describe the method of using string connected to a pencil to draw arcs or circles.

Answer: Connect the string to the pencil and then hold the string with your thumb (see fig. 12.15).

SKETCHING STRAIGHT-LINE FIGURES

In the practice of freehand drawing, the student should begin with the simplest figures before sketching the more difficult. The figures to be sketched may be classed as having either two or three di-

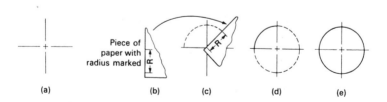

Piece of
paper with
radius marked

(a)   (b)   (c)   (d)   (e)

Fig. 12.14. Steps in sketching a circle.

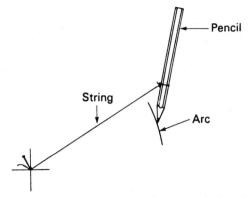

Fig. 12.15. A string connected to a pencil may be used to draw circles and arcs.

mensions. Two-dimensional figures have length and breadth only; three-dimensional figures have length, breadth, and thickness.

TWO-DIMENSIONAL FIGURES

The simplest example of a two-dimensional figure is a square or rectangle. To sketch a square, first locate two points, as A and B (fig. 12.16), at a horizontal distance apart equal to a side of the square. Place points C and D exactly under B and A and, at a distance equal to AB, make AD and AC equal to AB. Now, join AB and DC, then join AD and BC.

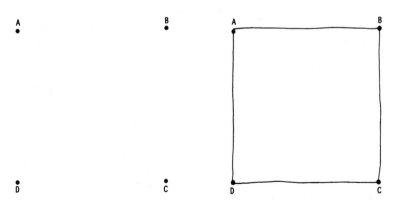

Fig. 12.16. Freehand method of drawing a square.

Evidently by joining the lines in the order of parallel pairs, less effort is required than if joined in such order as *AB*, *AD*, *DC*, and *BC*. Note in this order the position of the arm must be changed laterally four times. It is not the intention of the author to fill this chapter with a multiplicity of geometrical problems, but to give only a few typical figures to serve as practice examples in sketching.

Exercise—Find the center of a square.

Having drawn a square, as directed in figure 12.16, draw the two diagonals *AC* and *BD*. Point *O* where they intersect is the center of the square, as shown in figure 12.17.

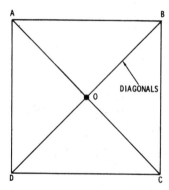

Fig. 12.17. Locating the center of a square by drawing diagonals.

ELLIPSES

An ellipse is a curved figure which the drafter must frequently sketch.

What is an ellipse?

> *Answer:* A projection of a circle on a plane oblique to that of the circle, as shown pictorially in figure 12.18.

What is the mathematical definition of an ellipse?

> *Answer:* An ellipse is a plane curve such that the sum of the distances from any point on the curve to two fixed points is a constant.

What are the two fixed points called?

> *Answer:* The foci.

176

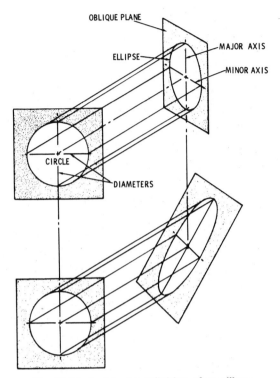

Fig. 12.18. A pictorial definition of an ellipse.

Name the two diameters of an ellipse.

Answer: The longest diameter of the ellipse is called the *major axis* and the shortest diameter the *minor axis*, as shown in figure 12.19.

What is the relation between the major and minor axes of an ellipse?

Answer: They bisect each other at right angles at a point called the center.

What are the two limits used to form an ellipse?

Answer: Its contour may vary from a straight line to a circle.

Define the diameter of an ellipse.

Answer: The diameter of an ellipse is any line drawn

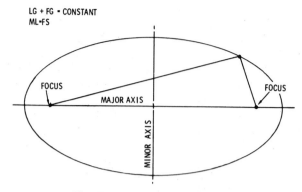

Fig. 12.19. Illustrating an ellipse.

through the center and terminated by the curve at each extremity.

What are conjugal diameters?

Answer: Two diameters so related that the tangents at the ends of either are parallel to the other.

Ellipses can be drawn in infinite variety, as to length and width. The major and minor axes may have any imaginable difference in length. By holding a penny in a vertical position, and then gradually turning it until hardly any of the surface is visible, the circular edge of the coin presents every conceivable change in the form of an ellipse.

Name some methods of describing ellipses.

Answer: Freehand, semifreehand, and plotting points.

Freehand Ellipses

Attempts to describe an ellipse by the preliminary movement method usually results in figures more distorted than elliptical. A good drafter, though, can describe an ellipse with some accuracy by this method.

How is an ellipse described by the freehand method?

Answer: Draw, as shown in figure 12.20, a rectangular axes XX and YY intersecting at O. By eye, mark off the major and minor axes AB and CD, making $OA = OB$, and $OC = OD$. Then, by the preliminary movement method (already described), using elliptical instead of circular motion, try to ap-

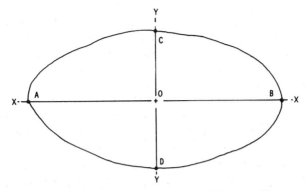

Fig. 12.20. Freehand method of describing an ellipse.

proximate the ellipse, as shown in figure 12.20. The difficulty is apparent by the distorted figure produced. This method is better adapted to small ellipses without trying to describe about their axes.

Semifreehand Ellipses

This is, strictly speaking, a freehand method, but a series of guide points are first located by the eye, and the curve described by progressive short strokes.

Describe the semifreehand method.

Answer: First, draw the axes as shown in figure 12.21; then locate by eye a series of guiding points, such as 1, 2, 3, *a*, *b*, *c*, etc., as judged by eye to be on the curve. The best

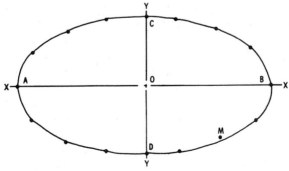

Fig. 12.21. Illustrating a semifreehand method of describing an ellipse.

179

way to get these points near the correct position is to see the entire curve mentally.

The final step is to describe, with short strokes, a curve through these points. Any point that is evidently off, as at *M*, fair up the curve by drawing around it either outside or inside.

Ellipses With Plotted Points

This is a semimechanical method in which points lying on the curve are obtained mechanically and the curve described through these points freehand. A very close approximation can be obtained by this method.

Describe the method with plotted points.

Answer: First, draw the rectangular axes *XX* and *YY*, as shown in figure 12.22. Lay off *OA* and *OC* half major and minor axes, respectively. It is not necessary to lay off the other halves of these axes, since they will be automatically located in plotting more accurately than by eye.

On a strip of paper, mark *OA* = ½ major axis and *OC* = ½ minor axis, as at *L*. Place the strip of paper in progressive positions so that the point *C* is always on the major axis and

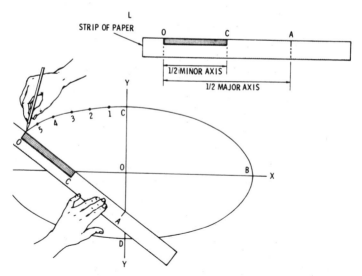

Fig. 12.22. Semimechanical method of describing an ellipse by plotting points on a curve.

point *A* on the minor axis, making a dot for each position, as 1, 2, 3, etc. Through the points thus obtained, sketch in the curve by short progressive strokes. Repeat for each quadrant, thus completing the ellipse. Figure 12.23 shows progressive positions of the strip of paper for the first quadrant.

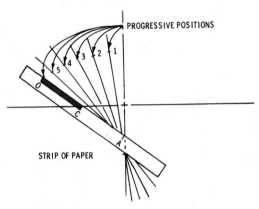

Fig. 12.23. Progressive positions of the strip of paper for the first quadrant in describing an ellipse by plotting.

THREE-DIMENSIONAL FIGURES

The term *three dimension* relates to figures having length, breadth, and thickness.

What is the simplest example of a three-dimensional figure?

Answer: A cube. In a cube, any of the three dimensions may be called length, breadth, or thickness, but in elongated figures these terms are more specific.

What does the first attempt to sketch a cube in cabinet projection look like?

Answer: Something like the distorted figure shown in figure 12.24.

Why?

Answer: Because little attention is paid to the direction of the axes, which is the basis for drawing the cube.

Exercise—Draw in cabinet projection the cube shown in figure 12.25.

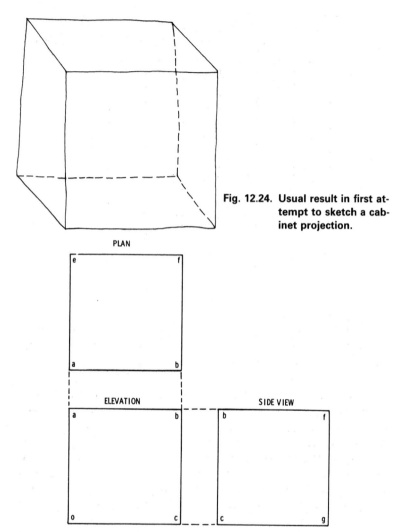

Fig. 12.24. Usual result in first attempt to sketch a cabinet projection.

PLAN

e   f

a   b

ELEVATION   SIDE VIEW

a   b   b   f

o   c   c   g

Fig. 12.25. Orthographic views of a cube.

First, draw the axes OX, CY, and OZ, as shown in figure 12.26. Now, lay off the points a and c on these axes, so that Oa and Oc will each equal a side of the cube. On the OZ axis, lay off $Oh = \frac{1}{2}$ of Oa. Draw ab parallel to Oc and join bc, Similarly, draw ae, bf, and cg parallel to OZ, and ef and hg parallel to OX, thus completing the square.

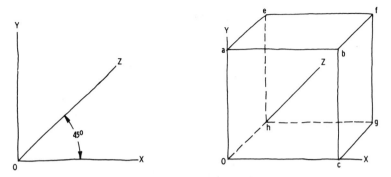

Fig. 12.26. A method of drawing a cube freehand in cabinet projection.

What is the important thing to remember?

 Answer: Parallelism. Axis lines *XX* must be parallel to *OX*, axis *YY* lines parallel to *OY*, etc.

Exercise—Draw a cube in isometric projection. (figure 12.27).

 The same method is used as for cabinet projection, except that the axes are at different angles, and there is no foreshortening in any of the three planes.

What is the objection to the drawings shown in figure 12.27?

 Answer: Since no projection plane lies in the plane of the paper, it is necessary to construct ellipses to represent circular portions of an object.

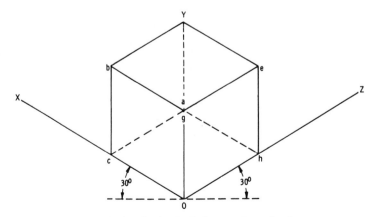

Fig. 12.27. A sketch of cube in isometric projection.

183

Exercise—Make a sketch of a cylinder end showing the head, stuffing box, and piston rod, as in figure 12.28.

First, draw two axes, as *XX* and *YY*. Evidently for a symmetrical figure like this, it is not necessary to show all in section. Sketch the part in sectional view at the right of *YY*, and in full view at the left of *YY* The horizontal axis here serves well for the dividing line between the cylinder and head. One part of the stuffing box is shown solid black to avoid too much angularity of section lines.

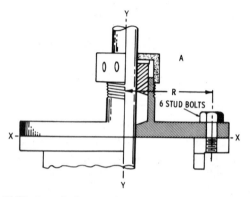

Fig. 12.28. A typical example of sketching machine details.

How are such objects as the piston rod and bolts or studs shown?
    Answer: In full view.

When should sketches be dimensioned?
    Answer: Sketches are intended simply to record an idea, as in the development of some machine. If the sketch is intended for shop use, it should have all dimensions and any other data necessary to produce the object.

CHAPTER 13

Map and Topographical Drawing

This book would not be complete without a chapter on maps and topographical drawing. This subject comes chiefly under surveying, but all building structures must be tied into a survey and topography. It is highly important in the layout of a building that the building itself be level, and the land around it must drain and fit into the picture of the surrounding area. Without this, flood problems will be created, and laws and ordinances may be ignored. In the construction of a building, one must not interfere with the rights of adjoining land owners.

This is accomplished by surveys to establish the contours of the ground, and the establishment of the property line must be referred to known and established land marks. These are usually established section corners. The basis of establishing all of this is by use of United States Geological Survey (USGS) maps. These maps provide a common base for starting. The elevations are established from a known sea-level figure and there are established bench marks throughout the nation that are used.

USGS maps show townships, sections, ranges, and base lines, and make corrections for changes in longitude and latitude. They show the important established features, such as rivers, lakes, ditches, swamps, roads, and other pertinent information. The mechanic is not usually going to have to know all that pertains to a survey, because surveys are made by registered surveyors, but the mechanic has to interpret and understand the surveys. In this manner he comes to know whether the building is being placed in the right location in respect to the subdivision, city lot and block, or farm in the country.

A *plat* is a map with the third dimension omitted. A plat of a survey sets forth all of the necessary information for a legal description of the land covered, and should contain the following items:

1. Lengths and bearings of the several sides
2. Acreage
3. Location and description of monuments found and set
4. Locations of streams, highways, etc.
5. Names of owners of adjacent property
6. Title and north point
7. Certification
8. Official division lines with the tract of land
9. Contours to indicate the topography, if required

What is the most common unit of measurement used?
  Answer: The foot.

How many square feet are there in an acre?
  Answer: 43,560 square feet per acre.

How many feet in a mile?
  Answer: 5,280 feet.

How many acres in a square mile?
  Answer: 640 acres.

How large is a section of land?
  Answer: It is normally considered to be one square mile. Because of correction lines that are necessary to correct for longitude and latitude, a section will not always be a true square mile. But ordinarily it is thought of as one square mile.

What is a township?
  Answer: A township is six miles wide, six miles long, and contains approximately 36 square miles.

What is a base line?
  Answer: A base line is an east and west line that has been established as a starting point for townships. Townships are indicated in respect to the base line, such as north or south of the base line. Thus, a township immediately north of the base line would be township 1 north, and a township immediately south of a base line would be township 1 south.

How are townships located east and west?
  Answer: Principal meridians have been established, and

thus townships are located east or west from the meridian line. For instance, the first township west of the 6th principal meridian would be range 1 west of the 6th P.M.

How can you find the base lines and meridian lines?

Answer: These have been established nationally on maps; thus, one may readily establish the area that he wants from these maps.

Draw a township 4 north, range 67 west of the 6th P.M., indicating the section numbers, etc., as shown in figure 13.1.

Ordinarily, the north indicator on a plat is pointed to the true north. There are two norths, the *true north* and the *magnetic north.* The true north stays in a fixed location, but the magnetic north will vary. The magnetic north is shown on an *isogonic chart,* which will have to be corrected from time to time as the magnetic north varies back and forth. The point of 0 (zero declination between the true north and the magnetic north) is known as the *agonic line.* In maps, the true north is used because it does not vary from year to year, and corrections for the magnetic north may be made by taking into consideration the declination at the time of the survey.

Figure 13.2 shows the true meridian or true north, the magnetic meridian or magnetic north, and the east and west line in relation

R 67W OF 6TH P.M.

6	5	4	3	2	1
7	8	9	10	11	12
18	17	16	15	14	13
19	20	21	22	23	24
30	29	28	27	26	25
31	32	33	34	35	36

T 4 N

N

Fig. 13.1. Illustrating a plot of land called a township, which is divided into one mile sections.

187

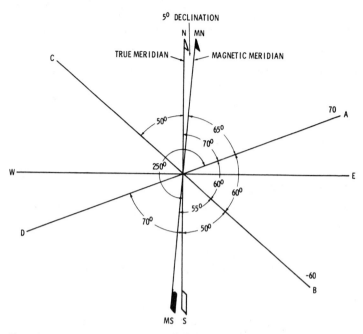

Fig. 13.2. Illustrating true north, magnetic meridian, or magnetic north.

to the true north. We would describe the location of the magnetic north as N 05°00′00″E, which would mean that the magnetic north is 5° east of north, line A is N 70°00′00″E, which means line A is 70° east of north. Line B is S 50°00′00″E, which means line B is in a southerly direction and 50° east of south. Line C is N 50°00′00″W, meaning that line C is northerly and 50° west of north. Line D is S 70°00′00″W, meaning line D is southerly 70° west of south.

The above explanation of bearings is given as it is used on land maps, subdivisions, city plats, etc. In checking surveys and plats, the magnetic declination must be known or established for the time that you are making the survey check. There are usually known established monuments or pins from which one may get the necessary information. Due to the fact that the earth is round, corrections must be made for this curvature. Figure 13.3 has an exaggerated curvature to illustrate more clearly the corrections that will be necessary. Descriptions of tracts of lands are described in relationship to sections and parts of sections. With a little study, this will become very clear to you, as the abbreviations will guide you exactly to the parcel of land that is being described in the area map.

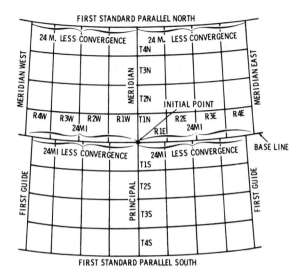

Fig. 13.3. The correction lines due to the curvature of the earth.

Figure 13.4 illustrates sections divided into lots, such as section 5, township 3 north, range 67 west of the 6th principal meridian. This is the part indicated by *ABDC*, and would be abbreviated as sec. 5, T3 N, R67 W of 6th P.M. (including county and state);

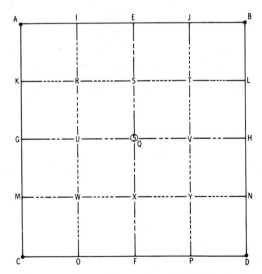

Fig. 13.4. A one-mile section divided into lots.

EBHQ is the NE ¼ sec. 5, T3N,R67W; *AEQG* is the NW ¼ sec. 5, etc.; *GQFC* is the SW ¼ sec. 5, etc.; *QHDF* is the SE ¼ sec. 5, etc.; *EBDF* is the E ½ sec. 5, etc.; *EJPF* is the W ½, E ½, sec. 5, etc.; *MNDC* is the S ½, S ½ sec. 5, etc.; *JBLT* is the NE ¼, NE ¼ sec. 5, etc.; *STVQ* is the SW ¼, NE ¼ sec. 5, etc.; *D* is the SE cor. sec. 5, etc.; *C* is the SW cor. sec. 5, etc.; *A* is the NW cor. NW ¼ sec. 5, etc.

From the above descriptions, you will obtain an idea of how to locate parts of a section. Spend a little time on this, as it is important to be able to identify the part of a section one is talking about. Zoning is becoming universal in our land; it is made necessary by the crowding together of people due to population growth. Land values must be maintained and adjacent land owners must not have the value of their land affected by structures on adjacent land.

It seems that there must be a control placed on development and plans made for the future. Natural resources and agricultural land must be protected. One of our most valuable assets is water, which must also be protected. Major cities will be out of gravel in a few years, even though there is an estimated 70 million tons of the precious item in the vicinity. This gravel is not available, how-ever, because planning was not started early enough, and the land under which the gravel lies was built upon with buildings and other structures.

You might wonder what this has to do with blueprint reading, but in building, the zoning must first be considered; from there we expand out to the subdivisions, building sites, etc. Figure 13.5 illustrates a small zoning area map used as a guide for subdividing into lots.

In engineering, construction, and subdivisions layout, and for many other purposes, it will be necessary to lay out a plat. The topography shows the land elevations at the contour intervals that are appropriate for the job at hand. In geological surveys, 10-foot contour intervals are often used; in building and subdivisions, 1- or 2-foot contour intervals might be required. These are necessary to give a better picture of drainage, flooding, and how much dirt has to be removed or added to bring the ground to the right elevations for proper use.

Figure 13.6 illustrates a preliminary plat for a subdivision. In this illustration you will see that the broken and dashed lines are the contours. The broken line indicates multiples of ten, the dashed lines are 2-foot intervals in between the multiples of 10. Shown in

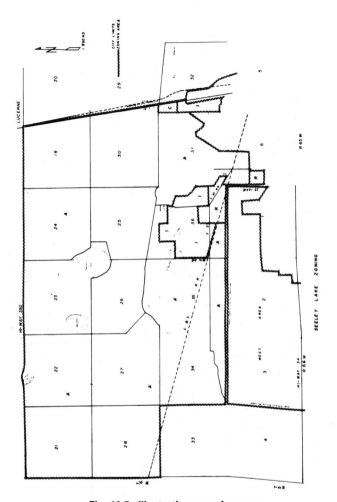

Fig. 13.5. Illustrating a zoning map.

this plat are the adjacent land owners, easements for utilities, and drainage. Note the lack of bearings on this plat. The reason for this is that if the plat should not be approved, the money spent in figuring the bearings would be lost. They will appear on the final plat. Note the contour lines and you will see exactly how the surface water will drain, which is an important factor when assigning a value to the land.

The subdivider and builders can tell how much soil will have to be moved, added, or taken away to prepare the land for proper

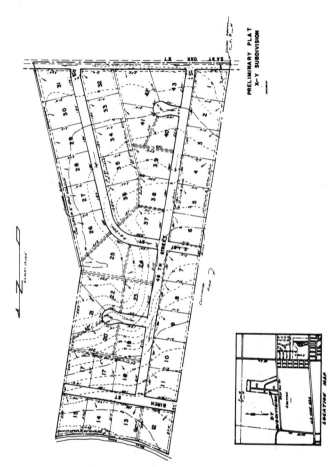

Fig. 13.6. Illustrating a preliminary plat for a subdivision.

building. This is also important in securing loans for buildings and homes. The contour elevations are taken from a known elevation, such as a geodetic bench mark, in the vicinity.

What are contours on a map?

> *Answer:* They are lines that indicate the elevations of the land.

On what intervals are they taken?

> *Answer:* The contour intervals depends entirely upon their use. They may be at 6″, 1′, 2′, 5′, 10′, or any other interval that is practical for the application.

192

What is a bearing?

Answer: A bearing shows the angle of departure from the four principal points of the compass. (The word compass is used here to indicate true north, south, east, and west.)

A piece of property 300' by 200' is surveyed by establishing a grid 100' on each side. The elevations of these points are given below. This field is to be leveled without the use of borrow or fill material. What would be the new elevation of the field? Assume no shrinkage or swell of the material.

A — 1 = 92.0'	B — 1 = 96.0'	C — 1 = 100.0'
A — 2 = 90.0'	B — 2 = 92.0'	C — 2 = 94.0'
A — 3 = 87.0'	B — 3 = 89.0'	C — 3 = 90.0'
A — 4 = 84.0'	B — 4 = 88.0'	C — 4 = 89.0'

Answer: This is solved by the averages method. The elevation of each corner is multiplied by the number of corners at that point. Thus, A-4 is one corner, B-4 is two corners, B-3 is four corners, A-3 is two corners, and C-3 is two corners, as shown in figure 13.7.

$$
\begin{array}{rcr}
84 \times 1 &=& 84 \\
88 \times 2 &=& 176 \\
89 \times 1 &=& 89 \\
87 \times 2 &=& 174 \\
89 \times 4 &=& 356 \\
90 \times 2 &=& 180 \\
90 \times 2 &=& 180 \\
92 \times 4 &=& 368 \\
94 \times 2 &=& 188 \\
92 \times 1 &=& 92 \\
96 \times 2 &=& 192 \\
100 \times 1 &=& 100 \\
\hline
24 && 2179
\end{array}
$$

From these figures, we find 24 corners and a total of 2179', so we divide 2179 by 24 and obtain an average of 90.79166', which we will round off to 90.79'. This will be the new elevation of the land leveled.

In the preceding question, we want an elevation of 95' over the

entire area. How many cubic yards of material will have to
be hauled in?

Answer:

$$95 - 90.79 = 4.21 \text{ feet.}$$

$$300 \times 200 = 60,000 \text{ sq. ft.}$$

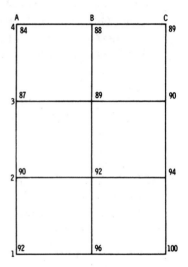

Fig. 13.7. Diagram for land leveling.

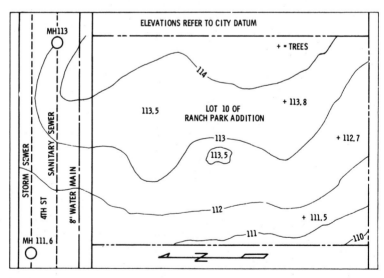

Fig. 13.8. Illustrating contours of land.

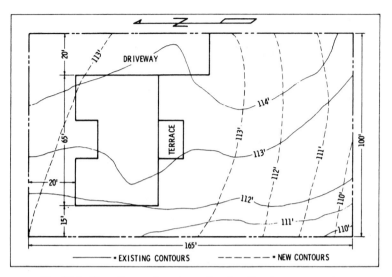

Fig. 13.9. Elevation and contour after leveling for construction.

$$60,000 \times 4.21 = 252,600 \text{ cu. ft.}$$

$$27 \text{ cu. ft.} = 1 \text{ cu. yard}$$

$$252,600 \text{ cu. ft. divided by } 27 = 9355.5 \text{ cubic yards of fill}$$

Figure 13.8 illustrates a plot of ground on which a building is to be placed. The contours and elevations as they exist are shown, along with the storm sewer, sanitary sewer, and water lines. Figure 13.8, shows lot 10 as it exists. Figure 13.9 shows both the existing elevations and the contours as they will be after leveling, as well as the intended construction. The street, sewers, etc., are not shown on this plat, merely the property lines. The plot or lot of land will be surveyed, stakes placed, and the land leveled to the new contours. If there is to be a basement, this will be dug, and, if necessary, the dirt from the excavation may be used for fill.

CHAPTER 14

Graphic Symbols

In drafting and blueprint work, a form of shorthand is used to illustrate what is to be installed and at what point or location the installation will be in the building. These are commonly known as *graphic symbols*. These will be given here in sections, and proper recognition will be given to the organizations that have so kindly given permission to reproduce these symbols. The ANSI bulletin number will also be given.

Various organizations establish the standard symbols, and they are approved by the American National Standards Institute and printed under a number assigned to the particular set. The following abbreviations appear below:

American National Standards Institute (ANSI)

Institute of Electrical and Electronics Engineers (IEEE)

Institute of Radio Engineers (IRE)

American Institute of Electrical Engineers (AIEE)

American Society of Mechanical Engineers (ASME)

At one time the IRE and the AIEE were separate, but recently they merged as the IEEE. The first section of this chapter will be devoted to what is known as the ANSI, Y32 series, covering electrical and electronic symbols.[1] The symbols that appear in the ANSI booklets have been compiled by numerous organizations and many dedicated men. Direct quotes from the explanatory material will appear in this book in *italics*.

---

1. The author wishes to thank and acknowledge the permission granted by the Institute of Electrical and Electronics Engineers, 345 East 47th Street, New York, NY, 10017, through Mr. John J. Anderson, Secretary, IEEE Standards committee, and to the American National Standards Institute for the use of these symbols in this book.

The symbols shown are standards for industry. However, some designers or individual institutions deviate from these standards. Where this is done, a legend showing what the symbols mean should be added to the prints, as the industries are quite well established on the standard symbols shown in this book.

GRAPHIC ELECTRICAL WIRING SYMBOLS

American Standard Graphic electrical wiring symbols for architectural and electrical layout drawings.

0.1 DRAFTING PRACTICES *applicable to graphic electrical wiring symbols.*

  a. *Electrical layouts should be drawn to an appropriate scale or figure-dimensions noted. They should be made on drawing sheets separate from the architectural or structural drawings or the drawing sheets for mechanical or other facilities.*

  Clearness of drawings is often reduced when all different electric systems to be installed in the same building area are laid out on the same drawing sheet. Clearness is further reduced when an extremely small drawing scale is used. Under these circumstances, each or certain of the different systems should be laid out on separate drawing sheets. For example, it may be better to show signal system outlets and circuits on drawings separate from the lighting and power branch circuit wiring.

  b. *Outlet and equipment locations with respect to the building should be shown as accurately as possible on the electrical drawing sheets to reduce reference to architectural drawings. Where extremely accurate final location of outlets and equipment is required, figure dimensions should be noted on the drawings. Circuit and feeder run lines should be so drawn as to show their installed location in relation to the building insofar as it is practical to do so. The number and size of conductors in the runs should be identified by notation when the circuit run symbol does not identify them.*

  c. *All branch circuits, control circuits and signal system circuits should be laid out in complete detail on the electri-*

cal drawings including identification of the number, size and type of all conductors.

d. Electric wiring required in conjunction with such mechanical facilities as heating, ventilating and air conditioning equipment, machinery and processing equipment should be included in detail in the electrical layout insofar as possible when its installation will be required under the electrical contract. This is desirable to make reference to mechanical drawings unnecessary and to avoid confusion as to responsibility for the installation of the work.

e. A complete electrical layout should include at least the following on one or more drawings:

1. Floor plan layout to scale of all outlet and equipment locations and wiring runs.

2. A complete schedule of all of the symbols used with appropriate description of the requirements.

3. Riser diagram showing the physical relationship of the service, feeder and major power runs, units substations, isolated power transformers, switchboards, panel boards, pull boxes, terminal cabinets and other systems and equipment.

4. Where necessary for clearness, a single line diagram showing the electrical relationship of the component items and sections of the wiring system.

5. Where necessary to provide adequate information elevations, sections and details of equipment and special installations and details of special lighting fixtures and devices.

6. Sections of the building or elevation of the structure showing floor to floor, outlet and equipment heights, relation to the established grade, general type of building construction, etc. Where practicable, suspended ceiling heights indicated by figure dimensions on either the electrical floor plan layout drawings or on the electrical building section or elevation drawings.

7. Where necessary to provide adequate information plot plan to scale, showing the relation of the building or structure to other buildings or structures, service

poles, service manholes, exterior area lighting. exterior wiring runs, etc.

8. In the case of exterior wiring systems for street and highway lighting, area drawings showing the complete system.

9. Any changes to the electrical layout should be clearly identified on the drawings when such changes are made after the original drawings have been completed and identified on the drawing by a revision symbol. △

0.2 EXPLANATION SUPPLEMENTING THE SCHEDULE OF SYMBOLS

a. GENERAL

1. TYPE OF WIRING METHOD OR MATERIAL REQUIREMENT: *When the general wiring method and material requirements for the entire installation are described in the specifications or specification notations on drawings, no special notation need be made in relation to symbols on the drawing layout, e.g., if an entire installation is required by the specifications and general reference on the drawings to be explosion proof, the outlet symbols do not need to have special identification.*

    When certain different wiring methods or special materials will be required in different areas of the building or for certain sections of the wiring system or certain outlets, such requirements should be clearly identified on the drawing layout by special identification of outlet symbols rather than only by reference in the specifications.

2. SPECIAL IDENTIFICATION OF OUTLETS: *Weather proof, vapor tight, water tight, rain tight, dust tight, explosion proof, grounded or recessed outlets or other special identification may be indicated by the use of uppercase letter abbreviations at the standard outlet symbol, e.g.:*

Weather proof	WP
Vapor tight	VT
Water tight	WT
Rain tight	RT
Dust tight	DT
Explosion proof	EP

Grounded	G
Recessed	R

The grade, rating and function of wiring devices used at special outlets should be indicated by abbreviated notation at the outlet location.

When the standard Special Purpose Outlet symbol is used to denote the location of special equipment or outlets or points of connection for such equipment, the specific usage will be identified by the use of a subscript numeral or letter alongside the symbol. The usage indicated by different subscripts will be noted on the drawing schedule of symbols.

b. *LIGHTING OUTLETS*

1. IDENTIFICATION OF TYPE OF INSTALLATION: *A major variation in the type of outlet box, outlet supporting means, wiring system arrangement and outlet connection and need of special items such as plaster rings or roughing-in cans, often depends upon whether a lighting fixture is to be recessed or surface mounted. A means of readily differentiating between such situations on drawings has been deemed necessary. In the case of a recessed fixture installation the adopted standard consists of a capital letter R drawn within the outlet symbol.*

2. FIXTURE IDENTIFICATION: *Lighting fixtures are identified as to type and size by the use of an upper-case letter, placed alongside each outlet symbol, together with a notation of the lamp size and number of lamps per fixture unit when two or more lamps per unit are required. A description of the fixture identified by the letter will be given either in the drawing schedule of symbols, separate fixture schedule on the drawing or in the electrical specifications.*

3. SWITCHING OF OUTLETS: *When different lighting outlets within a given local area are to be controlled by separately located wall switches, the related switching will be indicated by the use of lower-case letters at the lighting and switch outlet locations.*

c. SIGNALING SYSTEMS

1. BASIC SYMBOLS: *Each different basic category of signalling system shall be represented by a distinguishing Basic Symbol. Every item of equipment or outlet comprising that category of system shall be identified by that basic symbol.*

2. IDENTIFICATION OF INDIVIDUAL ITEMS: *Different types of individual items of equipment or outlets indicated by a basic system symbol will be further identified by a numeral placed within the open system basic symbol. All such individual symbols used on the drawings shall be included on the drawing schedule of symbols.*

3. USE OF SYMBOLS: *Only the basic signaling system outlet symbols are included in this Standard. The system or schedule of numbers referred to in (2) above will be developed by the designer.*

4. RESIDENTIAL SYMBOLS: *Signaling system symbols for use in identifying certain specific standardized residential type signal system items on residential drawing are included in this Standard. The reason for this specific group of symbols is that a descriptive symbol list such as is necessary for the above group of basic system symbols is often not included on residential drawings.*

d. POWER EQUIPMENT

1. ROTATING EQUIPMENT: *At motor and generator locations, note on the drawing adjacent to the symbol the horsepower of each motor, or the capacity of each generator. When motors and generators of more than one type or system characteristic, i.e., voltage and phase, are required on a given installation, the specific types and system characteristics should be noted at the outlet symbol.*

2. SWITCHBOARDS, POWER CONTROL CENTERS, UNIT SUBSTATIONS AND TRANSFORMER VAULTS: *The exact location of such equipment on the electrical layout floor plan drawing should be shown.*

   A detailed layout including plan, elevation and sectional views should be shown when needed for clearness showing the relationship of such equipment to the building structure or other sections of the electric system.

A single line diagram, using American Standard Graphic Symbols for Electrical Diagrams—Y32.2, should be included to show the electrical relationship of the components of the equipment to each other and to the other sections of the electric system.

e. *SYMBOLS NOT INCLUDED IN THIS STANDARD*

1. *Certain electrical symbols which are commonly used in making electrical system layouts on drawings are not included as part of this Standard for the reason that they have previously been included in American Standard Graphic Symbols for Electrical Diagrams, W32.2.*

   ASA policy requires that the same symbol not be included in two or more Standards. The reason for this is that if the same symbol were included in two or more Standards, when a symbol included in one Standard was revised, it might not be so revised in the other Standard at the same time, leading to confusion as to which was the proper symbol to use.

2. *Symbols falling into the above category include, but are not limited to, those shown below. The reference numbers are the American Standard Y32.2 item numbers.*

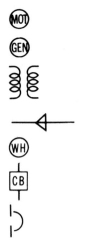

46.3 *Electric motor*

46.2 *Electric generator*

86.1 *Power transformer*

82.1 *Pothead (cable termination)*

48 *Electric watthour meter*

12.2 *Circuit element, e.g., circuit breaker*

11.1 *Circuit breaker*

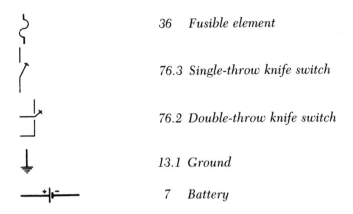

	36	Fusible element
	76.3	Single-throw knife switch
	76.2	Double-throw knife switch
	13.1	Ground
	7	Battery

LIST OF SYMBOLS

1.0 Lighting Outlets

Ceiling **Wall**

1.1 *Surface or pendant incandescent mercury vapor or similar lamp fixture*

1.2 *Recessed incandescent mercury vapor or similar lamp fixture*

1.3 *Surface or pendant individual fluorscent fixture*

1.4 *Recessed individual fluorescent fixture*

1.5 *Surface or pendant continuous-row fluorescent fixture*

1.6 *Recessed continuous-row fluorescent fixture*[2]

2. In the case of combination continuous-row fluorescent and incandescent spotlights, use combinations of the above standard symbols.

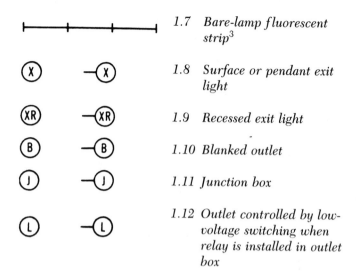

	1.7 Bare-lamp fluorescent strip[3]
	1.8 Surface or pendant exit light
	1.9 Recessed exit light
	1.10 Blanked outlet
	1.11 Junction box
	1.12 Outlet controlled by low-voltage switching when relay is installed in outlet box

2.0 Receptacle Outlets

Where all or a majority of receptacles in an installation are to be of the grounding type, the uppercase letter abbreviated notation may be omitted and the types of receptacles required noted in the drawing list of symbols and/or in the specifications. When this is done, any nongrounding receptacles may be so identified by notation at the outlet location.

Where weather proof, explosion proof or other specific types of devices are to be required, use the type of uppercase subscript letters referred to under Section 0.2 item a-2 of this Standard. For example, weather proof single or duplex receptacles would have the uppercase subscript letters noted alongside of the symbol.

Ungrounded   **Grounding**

2.1 *Single receptacle outlet*

2.2 *Duplex receptacle outlet*

3. In the case of continuous-row bare-lamp fluorscent strip above an area-wide diffusing means, show each fixture run, using the standard symbol; indicate area of diffusing means and type by light shading and/or drawing notation.

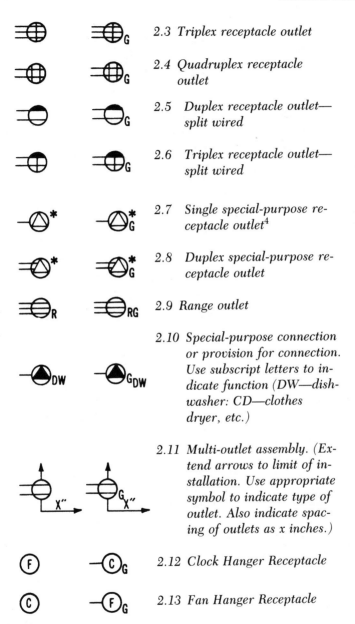

2.3 *Triplex receptacle outlet*

2.4 *Quadruplex receptacle outlet*

2.5 *Duplex receptacle outlet— split wired*

2.6 *Triplex receptacle outlet— split wired*

2.7 *Single special-purpose receptacle outlet[4]*

2.8 *Duplex special-purpose receptacle outlet*

2.9 *Range outlet*

2.10 *Special-purpose connection or provision for connection. Use subscript letters to indicate function (DW—dishwasher: CD—clothes dryer, etc.)*

2.11 *Multi-outlet assembly. (Extend arrows to limit of installation. Use appropriate symbol to indicate type of outlet. Also indicate spacing of outlets as x inches.)*

2.12 *Clock Hanger Receptacle*

2.13 *Fan Hanger Receptacle*

4. Use numeral or letter either within the symbol or as a subscript alongside the symbol keyed to explanation in the drawing list of symbols to indicate type of receptacle or usage. This applies also to 2.8 and 2.16.

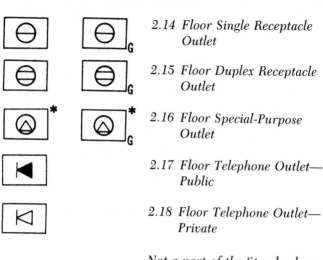

2.14 *Floor Single Receptacle Outlet*

2.15 *Floor Duplex Receptacle Outlet*

2.16 *Floor Special-Purpose Outlet*

2.17 *Floor Telephone Outlet— Public*

2.18 *Floor Telephone Outlet— Private*

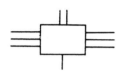

Not a part of the Standard: example of the use of several floor outlet symbols to identify a 2, 3, or more gang floor outlet

2.19 *Underfloor Duct and Junction Box for Triple, Double or Single Duct System as indicated by the number of parallel lines*

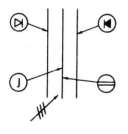

Not a part of the Standard: Example of use of various symbols to identify location of different types of outlets or connections for underfloor duct or cellular floor systems

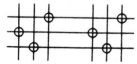

2.20 *Cellular Floor Header Duct*

3.0 Switch Outlets

S	*3.1*	*Single-pole switch*
S$_2$	*3.2*	*Double-pole switch*
S$_3$	*3.3*	*Three-way switch*
S$_4$	*3.4*	*Four-way switch*
S$_K$	*3.5*	*Key-operated switch*
S$_P$	*3.6*	*Switch and pilot lamp*
S$_L$	*3.7*	*Switch for low-voltage switching system*
S$_{LM}$	*3.8*	*Master switch for low-voltage switching system*

—⊖S *3.9 Switch and single receptacle*

=⊖S *3.10 Switch and double receptacle*

S$_D$	*3.11*	*Door switch*
S$_T$	*3.12*	*Time switch*
S$_{CB}$	*3.13*	*Circuit breaker switch*
S$_{MC}$	*3.14*	*Momentary contact switch or push-button for other than signaling system*
Ⓢ	*3.15*	*Ceiling pull switch*

SIGNALING SYSTEM OUTLETS

4.0 Institutional, Commercial, and Industrial Occupancies

Basic Symbol	Examples of Individual Item Identification (Not a part of the standard)	

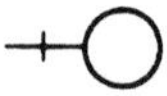

4.1 **I. Nurse Call System Devices (any type)**

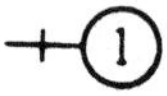

Nurses' Annunciator (can add a number after it is *24 to indicate number of lamps)*

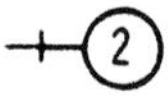

Call station, single cord, pilot light

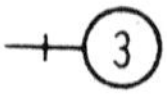

Call station, double cord, microphone-speaker

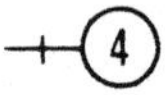

Corridor dome light, 1 lamp

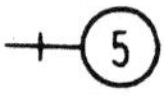

Transformer

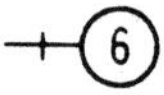

Any other item on same system—use numbers as required.

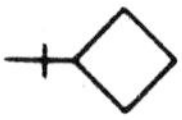

4.2 **II. Paging system Devices (any type)**

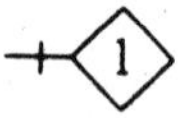

Keyboard

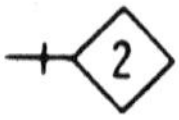

Flush annunciator

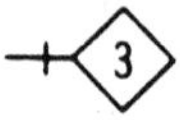

2-Face annunciator

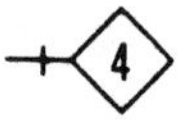

Any other item on same system—use numbers as required.

4.3 III. Fire Alarm System Devices (any type) including Smoke and Sprinkler Alarm Devices

Control panel

Station

10″ Gong

Pre-signal chime

Any other item on same system—use numbers as required.

4.4 IV. Staff Register System Devices (any type)

Phone operators' register

Entrance register—flush

Staff room register

Transformer

Any other item on same system—use numbers as required.

4.5 V. Electric Clock System Devices (any type)

Master clock

12″ Secondary—flush

12″ Double dial—wall mounted

209

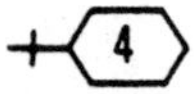

18″ Skeleton dial

Any other item on same system—use numbers as required.

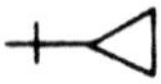

4.6 VI. Public Telephone System Devices

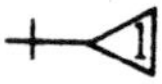

Switchboard

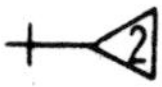

Wall phone

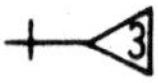

Any other item on same system—use numbers as required.

4.7 VII. Private Telephone System Devices (any type)

Switchboard

Desk phone

Any other item on same system—use numbers as required.

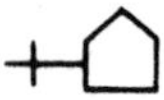

4.8 VIII. Watchman System Devices (any type)

Central station

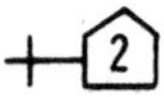

Key station

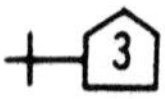

Any other item on same system—use numbers as required.

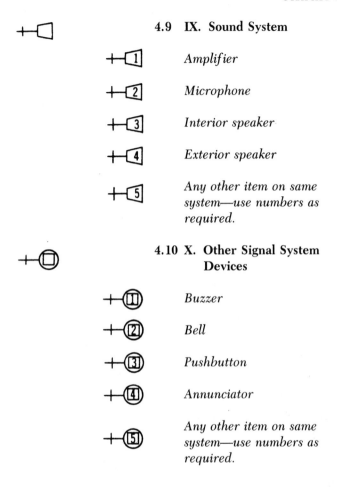

4.9   IX. Sound System

Amplifier

Microphone

Interior speaker

Exterior speaker

Any other item on same
system—use numbers as
required.

**4.10 X. Other Signal System
Devices**

Buzzer

Bell

Pushbutton

Annunciator

Any other item on same
system—use numbers as
required.

5.0 Residential Occupancies

*Signaling system symbols for use in identifying standardized resi-
dential-type signal system items on residential drawings where a
descriptive symbol list is not included on the drawing. When other
signal system items are to be identified, use the above basic symbols
for such items together with a descriptive symbol list.*

5.1   Pushbutton

5.2   Buzzer

⊏⊐ 5.3 *Bell*

⊏⊐✓ 5.4 *Combination bell-buzzer*

[CH] 5.5 *Chime*

◇— 5.6 *Annunicator*

[D] 5.7 *Electric door opener*

[M] 5.8 *Maid's signal plug*

☐ 5.9 *Interconnection box*

[BT] 5.10 *Bell-ringing transformer*

▶ 5.11 *Outside telephone*

▷ 5.12 *Interconnecting telephone*

[R] 5.13 *Radio outlet*

[TV] 5.14 *Television outlet*

6.0 Panelboards, Switchboards and Related Equipment

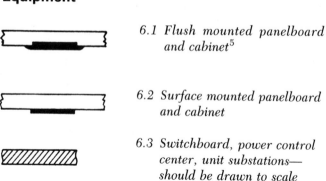

6.1 *Flush mounted panelboard and cabinet*[5]

6.2 *Surface mounted panelboard and cabinet*

6.3 *Switchboard, power control center, unit substations— should be drawn to scale*

5. Identify by notation or schedule. This applies to all numbers cited here except 6.6.

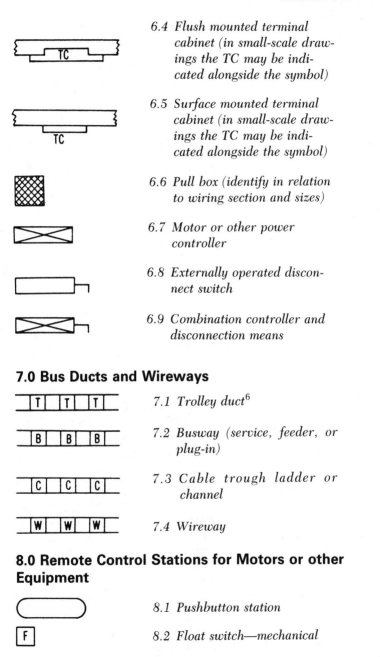

6.4 *Flush mounted terminal cabinet (in small-scale drawings the TC may be indicated alongside the symbol)*

6.5 *Surface mounted terminal cabinet (in small-scale drawings the TC may be indicated alongside the symbol)*

6.6 *Pull box (identify in relation to wiring section and sizes)*

6.7 *Motor or other power controller*

6.8 *Externally operated disconnect switch*

6.9 *Combination controller and disconnection means*

7.0 Bus Ducts and Wireways

7.1 *Trolley duct*[6]

7.2 *Busway (service, feeder, or plug-in)*

7.3 *Cable trough ladder or channel*

7.4 *Wireway*

8.0 Remote Control Stations for Motors or other Equipment

8.1 *Pushbutton station*

8.2 *Float switch—mechanical*

6. Identify by notation or schedule. This applies to all numbers in this section.

[L]   8.3 *Limit switch—mechanical*

[P]   8.4 *Pneumatic switch—
       mechanical*

[image: switch symbol with beam]   8.5 *Electric eye—beam source*

[image: switch symbol]   8.6 *Electric eye—relay*

—(T)   8.7 *Thermostat*

9.0 Circuiting

*Wiring method identification by notation on drawing or in
specification.*

—————————   9.1 *Wiring concealed in ceiling
            or wall*

— — — — — —   9.2 *Wiring concealed in floor*

- - - - - - - - - - - - -   9.3 *Wiring exposed*

*Note: Use heavy-weight line to
identify service and feed-
ers. Indicate empty con-
duit by notation CO (con-
duit only)*

                      2   1
————————→—→

                                  9.4 *Branch circuit home run to
                                  panel board. Number of ar-
                                  rows indicates number of
                                  circuits. (A numeral at each
                                  arrow may be used to iden-
                                  tify circuit number.) Note:
                                  Any circuit without further
                                  identifciation indicates two-
                                  wire circuit. For a greater
                                  number of wires, indicate
                                  with cross lines, e.g.:*

3 wires:

———/// ———

214

4 wires, etc.

Unless indicated otherwise, the wire size of the circuit is the minimum size required by the specification.

Identify different functions of wiring system, e.g., signaling system by notation or other means.

———————————o   9.5 Wiring turned up

———————————•   9.6 Wiring turned down

10.0 Electric Distribution or Lighting System, Underground

| M | 10.1 Manhole[7]

| H | 10.2 Handhole

| TM | 10.3 Transformer manhole or vault

| TP | 10.4 Transformer pad

   10.5 Underground direct burial cable (indicate type, size and number of conductors by notation or schedule)

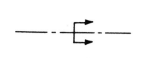   10.6 Underground duct line (indicate type, size, and number of ducts by cross-section identification of each run by notation or schedule. Indicate type, size, and number of conductors by notation or schedule)

7. Identify by notation or schedule. This applies to 10.1–4 and 10.7.

215

10.7 *Street light standard feed from underground circuit*

11.0 Electric Distribution or Lighting System Aerial

○   11.1 *Pole*[8]

    11.2 *Street light and bracket*

△   11.3 *Transformer*

───────   11.4 *Primary circuit*

------------   11.5 *Secondary circuit*

──────)   11.6 *Down guy*

──●───   11.7 *Head guy*

──○──)   11.8 *Sidewalk guy*

(───────   11.9 *Service weather head*

4 Arrester, Lightning Arrester (Electric surge, etc.) Gap

──● ●───   4.1 *General*

4.2 *Carbon block*
Block, telephone protector: The sides of the rectangle are to be approximately in the ratio of 1 to 2 and the space between rectangles shall be approximately equal to the width of a rectangle.

4.3 *Electrolytic or aluminum cell:*
This symbol is not composed of arrowheads.

8. Identify by notation or schedule. This applies to 11.1–5 and 11.9.

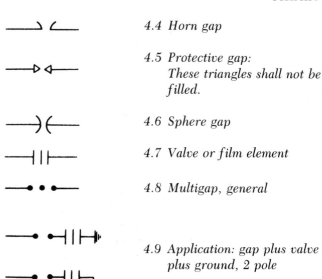

4.4 *Horn gap*

4.5 *Protective gap:*
    These triangles shall not be
    filled.

4.6 *Sphere gap*

4.7 *Valve or film element*

4.8 *Multigap, general*

4.9 *Application: gap plus valve*
    plus ground, 2 pole

7 Battery

The long line is always positive,
but polarity may be indicated in
addition.
Example:

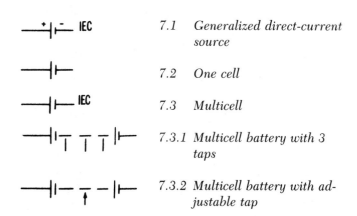

7.1   *Generalized direct-current*
    source

7.2   *One cell*

7.3   *Multicell*

7.3.1 *Multicell battery with 3*
    taps

7.3.2 *Multicell battery with ad-*
    justable tap

11 Circuit Breakers

If it is desired to show the condition causing the breaker to trip, the relay-protective-function symbols in item 66.6 may be used alongside the break symbol.

11.1   General

11.2   Air circuit breaker, if distinction is needed; for alternating-current breakers rated at 1,500 volts or less and for all direct-current circuit breakers

11.2.1 Network protector

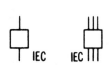

11.3   Circuit breaker, other than covered by item 11.2. The symbol in the right column is for a 3-pole breaker.

See note 11.3A

11.3.1 On a connection or wiring diagram, a 3-pole single-throw circuit breaker (with terminals shown) may be drawn as shown.

11.4   Applications

8. Note 11.3A—On a power diagram, the symbol may be used without other identification. On a composite drawing where confusion with the general circuit element symbol (item 12) may result, add the identifying letters CB inside or adjacent to the square. This applies also to 11.3.1.

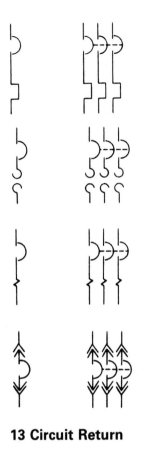

11.4.1 3-pole circuit breaker
with thermal overload
device in all 3 poles

11.4.2 3-pole circuit breaker
with magnetic overload
device in all 3 poles

11.4.3 3-pole circuit breaker,
drawout type

13 Circuit Return

13.1 Ground

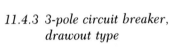

(A) A direct conducting connection to the earth or body of
water that is a part thereof.

(B) A conducting connection to a
structure that serves a function similar to that of an earth
ground (that is, a structure
such as the frame of an air,
space, or land vehicle that is
not conductively connected to
earth).

IEC

13.2 Chassis or frame connection
A conducting connection to a
chassis or frame of a unit. The
chassis or frame may be at a
substantial potential with
respect to the earth or structure
in which this chassis or frame
is mounted.

13.3 Common connections
Conducting connections made
to one another. All like-
designated points are connected.
The asterisk is not a part of
the symbol. Identifying values,
letters, numbers, or marks
shall replace the asterisk.

15 Coil, Magnetic Blowout

23 Contact, Electrical

For build-ups or forms using electrical contacts, see applications
under CONNECTOR (item 18), RELAY (item 66), SWITCH (item
76). See DRAFTING PRACTICES (item 0.4.6).

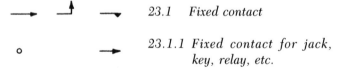

23.1   *Fixed contact*

23.1.1 *Fixed contact for jack,*
        key, relay, etc.

23.1.2 *Fixed contact for switch*

9. The broken line - — indicates where line connection to a symbol is made and is
not a part of the symbol.

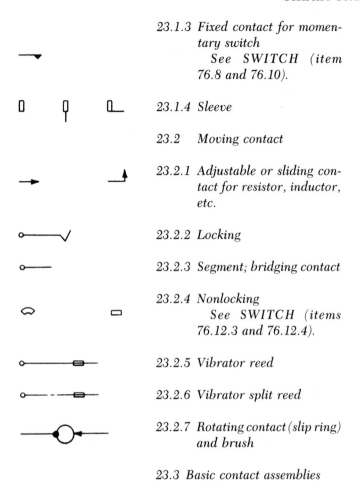

23.1.3 *Fixed contact for momentary switch*
See *SWITCH* (item 76.8 and 76.10).

23.1.4 *Sleeve*

23.2 *Moving contact*

23.2.1 *Adjustable or sliding contact for resistor, inductor, etc.*

23.2.2 *Locking*

23.2.3 *Segment; bridging contact*

23.2.4 *Nonlocking*
See *SWITCH* (items 76.12.3 and 76.12.4).

23.2.5 *Vibrator reed*

23.2.6 *Vibrator split reed*

23.2.7 *Rotating contact (slip ring) and brush*

23.3 *Basic contact assemblies*

The standard method of showing a contact is by a symbol indicating the circuit condition it produces when the actuating device is in the deenergized or nonoperated position. The actuating device may be of a mechanical, electrical, or other nature, and a clarifying note may be necessary with the symbol to explain the proper point at which the contact functions, for example, the point where a contact closes or opens as a function of changing pressure, level, flow, voltage, current, etc. In cases where it is desirable to show contacts in the energized or operated condition and where confusion may result, a clarifying note shall be added to the drawing. Auxiliary switches or contacts for circuit breakers, etc., may be designated as follows:

(a)   Closed when device is in energized or operated position.

(b)   Closed when device is in deenergized or nonoperated position.

(aa) Closed when operating mechanism of main device is in energized or operated position,

(bb) Closed when operating mechanism of main device is in deenergized or nonoperated position.

See American Standard C37.2-1962 for further details.

In the parallel-line contact symbols showing the length of the parallel lines shall be approximately 1 1/4 times the width of the gap (except for item 23.6)

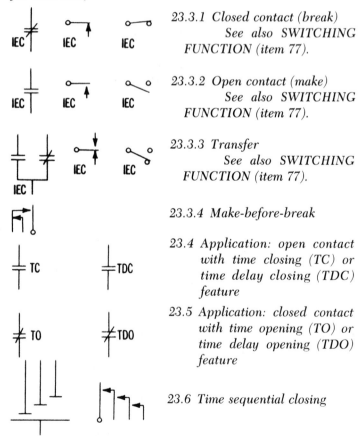

23.3.1 *Closed contact (break)*
   See also SWITCHING FUNCTION (item 77).

23.3.2 *Open contact (make)*
   See also SWITCHING FUNCTION (item 77).

23.3.3 *Transfer*
   See also SWITCHING FUNCTION (item 77).

23.3.4 *Make-before-break*

23.4 *Application: open contact with time closing (TC) or time delay closing (TDC) feature*

23.5 *Application: closed contact with time opening (TO) or time delay opening (TDO) feature*

23.6 *Time sequential closing*

24 Contactor.

See also RELAY (item 66).

Fundamental symbols for contacts, coils, mechanical connections, etc., are the basis of contactor symbols and should be used to represent contactors on complete diagrams. Complete diagrams of contactors consist of combinations of fundamental symbols for control coils, mechanical connections, etc., in such configurations as to represent the actual device.

Mechanical interlocking should be indicated by notes.

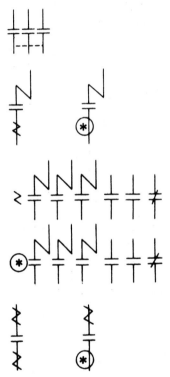

24.1 *Manually operated 3-pole contactor*

24.2 *Electrically operated 1-pole contactor with series blow-out coil*
   See note 24.2A[10]

24.3 *Electrically operated 3-pole contactor with series blow-out coils; 2 open and 1 closed auxiliary contacts (shown smaller than the main contacts)*

24.4 *Electrically operated 1-pole contactor with shunt blow-out coil*

46 Machine, Rotating[11]

46.1   *Basic*

10. Note 24.2A—The asterisk is not a part of the symbol. Always replace the asterisk by a device designation.

11. The information in note 10 applies to 46.5, 46.8–3, 46.8.5–7, 46.8.9–15, and 46.9.1–3.

46.2   *Generator (general)*

46.3   *Motor (general)*

46.4   *Motor, multispeed*
     USE BASIC MOTOR
     SYMBOL AND NOTE
     SPEEDS

46.5   *Rotating armature with
commutator and brushes*

46.6   *Field, generator or
motor.*
     *Either symbol of item
42.1 may be used in the
following items.*

46.6.1 *Compensating or commutating*

46.6.2 *Series*

46.6.3 *Shunt, or separately
excited*

46.6.4 *Magnet, permanent*
     See item 47.

46.7   *Winding symbols*
     *Motor and generator
winding symbols may be
shown in the basic circle
using the following
representation.*

46.7.1 *1-phase*

46.7.2 *2-phase*

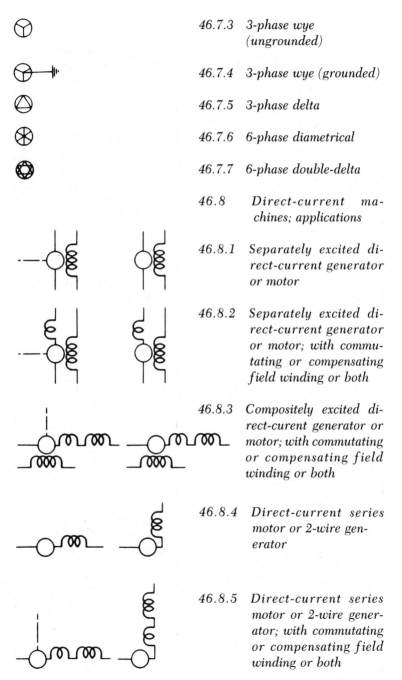

46.7.3 *3-phase wye (ungrounded)*

46.7.4 *3-phase wye (grounded)*

46.7.5 *3-phase delta*

46.7.6 *6-phase diametrical*

46.7.7 *6-phase double-delta*

46.8 *Direct-current machines; applications*

46.8.1 *Separately excited direct-current generator or motor*

46.8.2 *Separately excited direct-current generator or motor; with commutating or compensating field winding or both*

46.8.3 *Compositely excited direct-curent generator or motor; with commutating or compensating field winding or both*

46.8.4 *Direct-current series motor or 2-wire generator*

46.8.5 *Direct-current series motor or 2-wire generator; with commutating or compensating field winding or both*

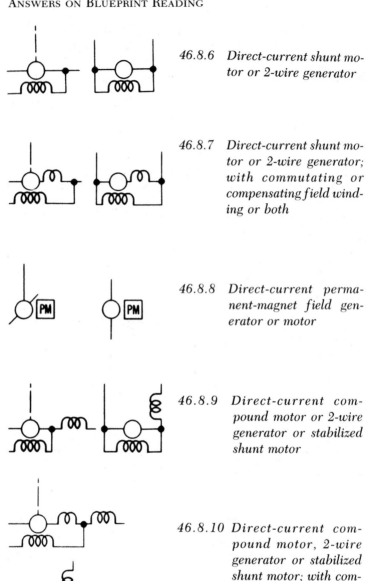

46.8.6   Direct-current shunt motor or 2-wire generator

46.8.7   Direct-current shunt motor or 2-wire generator; with commutating or compensating field winding or both

46.8.8   Direct-current permanent-magnet field generator or motor

46.8.9   Direct-current compound motor or 2-wire generator or stabilized shunt motor

46.8.10 Direct-current compound motor, 2-wire generator or stabilized shunt motor; with commutating or compensating field winding or both

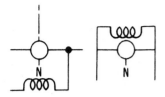

46.8.11 *Direct-current 3-wire shunt generator*

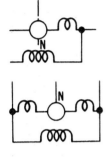

46.8.12 *Direct-current 3-wire shunt generator; with commutating or compensating field winding or both*

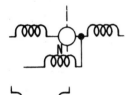

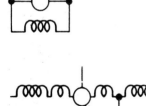

46.8.13 *Direct-current 3-wire compound generator*

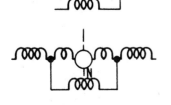

46.8.14 *Direct-current 3-wire compound generator; with commutating or compensating field winding or both*

227

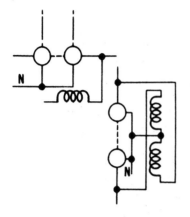

46.8.15 Direct-current balancer, shunt wound

46.9 Alternating-current machines; applicationss

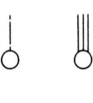

46.9.1 Squirrel-cage induction motor or generator, split-phase induction motor or generator, rotary phase converter, or repulsion motor

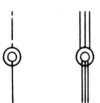

46.9.2 Wound-rotor induction motor, synchronous induction motor, induction generator, or induction frequency converter

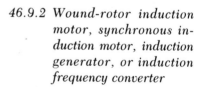

46.93 Alternating-current series motor

48 Meter Instrument

Note 48A—The asterisk is not a part of the symbol. Always replace the asterisk by one of the following letter combinations, depending on the function of the meter or instrument, unless some other identification is provided in the circle and explained on the diagram.

A	Ammeter IEC
AH	Ampere-hour
CMA	Contact-making (or breaking) ammeter
CMC	Contact-making (or breaking) clock
CMV	Contact-making (or breaking) voltmeter
CRO	Oscilloscope or cathode-ray oscillograph
dB	dB (decibel) meter
dBm	dBm (decibels referred to 1 milliwatt) meter
DM	Demand meter
DTR	Demand-totalizing relay
F	Frequency meter
G	Galvanometer
GD	Ground detector
I	Indicating
INT	Integrating
μA	Microammeter
mA	Milliammeter
NM	Noise meter
OHM	Ohmmeter
Op	Oil pressure
OSCG	Oscillograph string

PH	*Phasemeter*
PI	*Position indicator*
PF	*Power factor*
RD	*Recording demand meter*
REC	*Recording*
RF	*Reaction factor*
SY	*Synchroscope*
TLM	*Telemeter*
T	*Temperature meter*
THC	*Thermal converter*
TT	*Total time*
V	*Voltmeter*
VA	*Volt-ammeter*
VAR	*Varmeter*
VARH	*Varhour meter*
VI	*Volume indicator or meter, audio level*
VU	*Standard volume indicator or meter, audio level*
W	*Wattmeter*
WH	*Watthour meter*

———————— IEC

58.1 *Guided path, general*
A single line represents
the entire group of con-

ductors or the transmission path needed to guide the power or the signal. For coaxial and waveguide work, the recognition symbol is used at the beginning and end of each kind of transmission path and at intermediate points as needed for clarity. In waveguide work, mode may be indicated.

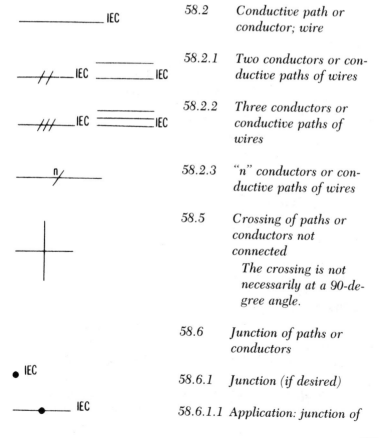

58.2   Conductive path or conductor; wire

58.2.1   Two conductors or conductive paths of wires

58.2.2   Three conductors or conductive paths of wires

58.2.3   "n" conductors or conductive paths of wires

58.5   Crossing of paths or conductors not connected

The crossing is not necessarily at a 90-degree angle.

58.6   Junction of paths or conductors

58.6.1   Junction (if desired)

58.6.1.1 Application: junction of

231

paths, conductor, or cable. If desired indicate path type, or size

58.6.1.2 Application: splice (if desired) of same size cables. Junction of conductors of same size or different size cables. If desired indicate sizes of conductors

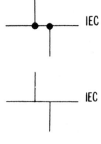

58.6.2 Junction of connected paths, conductors, or wires

OR ONLY IF REQUIRED
BY SPACE LIMITATION

63 Polarity Symbol

+ IEC   63.1 *Positive*

− IEC   63.2 *Negative*

76 Switch

See also FUSE (item 36); CONTACT, ELECTRIC (item 23); DRAFTING PRACTICES (items 0.4.6 and 0.4.7).

Fundamental symbols for contacts, mechanical connections, etc., may be used for switch symbols.

The standard method of showing switches is in a position with no operating force applied. For switches that may be in any one of two or more positions with no operating force applied and for switches actuated by some mechanical device (as in air-pressure, liquid-level, rate-of-flow, etc., switches), a clarifying note may be necessary to explain the point at which the switch functions.

When the basic switch symbols in items 76.1 through 76.4 are shown on a diagram in the closed position, terminals must be added for clarity.

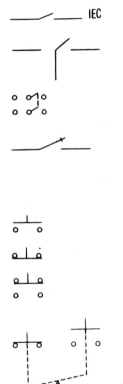

76.1   *Single throw, general*

76.2   *Double throw, general*

76.2.1 *Application: 2-pole dou-ble-throw switch with terminals shown*

76.3   *Knife switch, general*

76.6   *Pushbutton, momentary or spring return*

76.6.1 *Circuit closing (make)*

76.6.2 *Circuit opening (break)*

76.6.3 *Two-circuit*

76.7   *Pushbutton, maintained or not spring return*

76.7.1 *Two circuit*

86 Transformer

86.1 General

Either winding symbol may be used. In the following items, the left symbol is used. Additional windings may be shown or in di-

*cated by a note. For power trans-
formers use polarity marking H_1,
X_1, etc., from American Standard
C6.1-1956.*

For polarity markings on current and potential transformers, see items 86.16.1 and 86.17.1.

In coaxial and waveguide circuits, this symbol will represent a taper or step transformer without mode change.

	86.1.1	*Application: transformer with direct-current connections and mode suppression between two rectangular waveguides*
	86.2	*If it is desired especially to distinguish a magnetic-core transformer*
	86.2.1	*Application: shielded transformer with magnetic core shown*
	86.2.2	*Application: transformer with magnetic core shown and with a shield between windings. The shield is shown connected to the frame*
	86.6	*with taps, 1-phase*
	86.7	*Autotransformer, 1-phase*

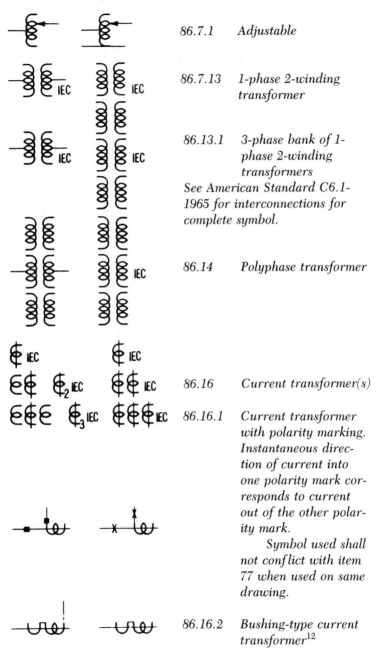

	86.7.1	Adjustable
	86.7.13	1-phase 2-winding transformer
	86.13.1	3-phase bank of 1-phase 2-winding transformers

See American Standard C6.1-1965 for interconnections for complete symbol.

86.14   Polyphase transformer

86.16   Current transformer(s)

86.16.1   Current transformer with polarity marking. Instantaneous direction of current into one polarity mark corresponds to current out of the other polarity mark.

Symbol used shall not conflict with item 77 when used on same drawing.

86.16.2   Bushing-type current transformer[12]

12. See note 9.

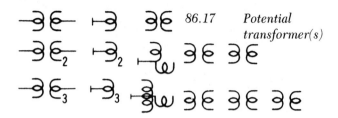

86.17 *Potential transformer(s)*

86.17.1 *Potential transformer with polarity mark. Instantaneous direction of current into one polarity mark corresponds to current out of the other polarity mark.*

Symbol used shall not conflict with item 77 when used on same drawing.

86.18 *Outdoor metering device*

86.19 *Transformer winding connection symbols*
For use adjacent to the symbols for the transformer windings.

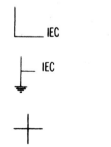

86.19.1 *2-phase 3-wire, ungrounded*

86.19.1.1 *2-phase 3-wire, grounded*

86.19.2 *2-phase 4-wire*

86.19.2.1	2-phase 5-wire, grounded
86.19.3	3-phase 3-wire, delta or mesh
86.19.3.1	3-phase 3-wire, delta, grounded
86.19.4	3-phase 4-wire, delta, ungrounded
86.19.4.1	3-phase 4-wire, delta, grounded
86.19.5	3-phase, open-delta
86.19.5.1	3-phase, open-delta, grounded at common point
86.19.5.2	3-phase, open-delta, grounded at middle point of one transformer
86.19.6	3-phase, broken-delta
86.19.7	3-phase, wye or star, ungrounded
86.19.7.1	3-phase, wye, grounded neutral The direction of the stroke representing the neutral can be arbitrarily chosen.
86.19.8	3-phase 4-wire, ungrounded

The following graphic symbols cover plumbing, heating, ventilation, and air conditioning.

ANSI Graphic Symbols For Plumbing ASA Y32.4-1955

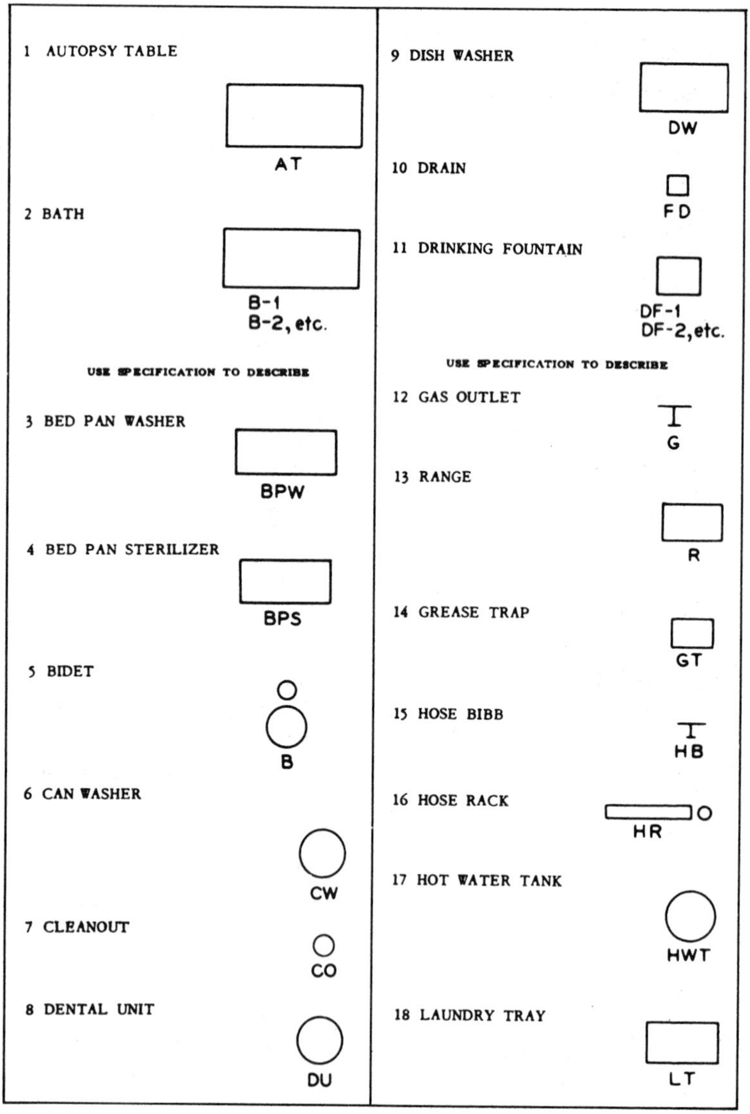

1 AUTOPSY TABLE

A T

2 BATH

B-1
B-2, etc.

USE SPECIFICATION TO DESCRIBE

3 BED PAN WASHER

BPW

4 BED PAN STERILIZER

BPS

5 BIDET

B

6 CAN WASHER

CW

7 CLEANOUT

CO

8 DENTAL UNIT

DU

9 DISH WASHER

DW

10 DRAIN

F D

11 DRINKING FOUNTAIN

DF-1
DF-2, etc.

USE SPECIFICATION TO DESCRIBE

12 GAS OUTLET

G

13 RANGE

R

14 GREASE TRAP

GT

15 HOSE BIBB

HB

16 HOSE RACK

HR

17 HOT WATER TANK

HWT

18 LAUNDRY TRAY

LT

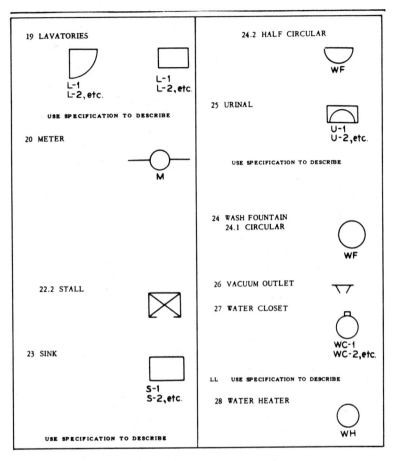

ANSI Graphic Symbols for Heating, Ventilating, and Air Conditioning ASA Z32.2.4-1949 (R1953)

AMERICAN NATIONAL STANDARD	
1 AIR ELIMINATOR	
2 ANCHOR	
3 EXPANSION JOINT	
4 HANGER OR SUPPORT	

5 HEAT EXCHANGER

6 HEAT TRANSFER SURFACE, PLAN
   (INDICATE TYPE SUCH AS CONVECTOR)

7 PUMP
   (INDICATE TYPE SUCH AS VACUUM)

8 STRAINER

9 TANK (DESIGNATE TYPE)

10 THERMOMETER

11 THERMOSTAT

12 TRAPS
    12.1 BOILER RETURN

    12.2 BLAST THERMOSTATIC

    12.3 FLOAT

    12.4 FLOAT AND THERMOSTATIC

    12.5 THERMOSTATIC

13 UNIT HEATER
    (CENTRIFUGAL FAN), PLAN

14 UNIT HEATER (PROPELLER), PLAN

15 UNIT VENTILATOR, PLAN

16 VALVES
    16.1 CHECK

    16.2 DIAPHRAGM

    16.3 GATE

    16.4 GLOBE

    16.5 LOCK AND SHIELD

    16.6 MOTOR OPERATED

    16.7 REDUCING PRESSURE

    16.8 RELIEF
          (EITHER PRESSURE OR VACUUM

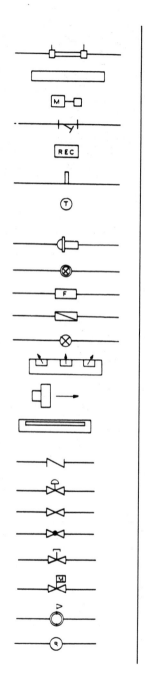

17 VENT POINT

18 ACCESS DOOR

19 ADJUSTABLE BLANK OFF

20 ADJUSTABLE PLAQUE

21 AUTOMATIC DAMPERS

22 CANVAS CONNECTIONS

23 DEFLECTING DAMPER

24 DIRECTION OF FLOW

25 DUCT (1ST FIGURE, SIDE SHOWN;
    2ND SIDE NOT SHOWN)

26 DUCT SECTION
    (EXHAUST OR RETURN)

27 DUCT SECTION (SUPPLY)

28 EXHAUST INLET CEILING
    (INDICATE TYPE

29 EXHAUST INLET WALL
    (INDICATE TYPE

30 FAN AND MOTOR WITH
    BELT GUARD

31 INCLINED DROP IN RESPECT
    TO AIR FLOW

32 INCLINED RISE iN RESPECT
    TO AIR FLOW

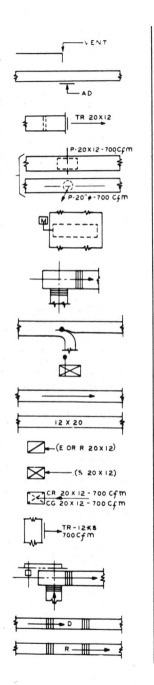

33 INTAKE LOUVERS ON SCREEN

34 LOUVER OPENING

35 SUPPLY OUTLET CEILING
   (INDICATE TYPE

36 SUPPLY OUTLET WALL
   (INDICATE TYPE

37 VANES

38 VOLUME DAMPER

39 CAPILLARY TUBE

40 COMPRESSOR

41 COMPRESSOR, ENCLOSED, CRANK-
   CASE, ROTARY, BELTED
42 COMPRESSOR, OPEN CRANKCASE,
   RECIPROCATING, BELTED
43 COMPRESSOR, OPEN CRANKCASE,
   RECIPROCATING, DIRECT DRIVE
44 CONDENSER, AIR COOLED,
   FINNED, FORCED AIR

45 CONDENSER, AIR COOLED,
   FINNED, STATIC

46 CONDENSER, WATER COOLED,
   CONCENTRIC TUBE IN A TUBE

47 CONDENSER, WATER COOLED,
   SHELL AND COIL

48 CONDENSER, WATER COOLED,
   SHELL AND TUBE

49 CONDENSING UNIT,
   AIR COOLED

50 CONDENSING UNIT,
   WATER COOLED

51 COOLING TOWER

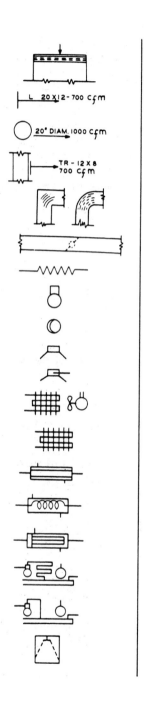

52 DRYER

53 EVAPORATIVE CONDENSER

54 EVAPORATOR, CIRCULAR, CEILING TYPE, FINNED

55 EVAPORATOR, MANIFOLDED, BARE TUBE, GRAVITY AIR

56 EVAPORATOR, MANIFOLDED, FINNED, FORCED AIR

57 EVAPORATOR, MANIFOLDED, FINNED, GRAVITY AIR

58 EVAPORATOR, PLATE COILS, HEADERED OR MANIFOLD

59 FILTER, LINE

60 FILTER & STRAINER, LINE

61 FINNED TYPE COOLING UNIT, NATURAL CONVECTION

62 FORCED CONVECTION COOLING UNIT

63 GAUGE

64 HIGH SIDE FLOAT

65 IMMERSION COOLING UNIT

66 LOW SIDE FLOAT

67 MOTOR-COMPRESSOR, ENCLOSED CRANKCASE, RECIPROCATING, DIRECT CONNECTED

68 MOTOR-COMPRESSOR, ENCLOSED CRANKCASE, ROTARY, DIRECT CONNECTED

69 MOTOR-COMPRESSOR, SEALED CRANKCASE, RECIPROCATING

70 MOTOR-COMPRESSOR, SEALED CRANKCASE, ROTARY

243

71 PRESSURESTAT

72 PRESSURE SWITCH

73 PRESSURE SWITCH WITH
    HIGH PRESSURE CUT-OUT

74 RECEIVER, HORIZONTAL

75 RECEIVER, VERTICAL

76 SCALE TRAP

77 SPRAY POND

78 THERMAL BULB

79 THERMOSTAT (REMOTE BULB)

80 VALVES

    80.1 AUTOMATIC EXPANSION

    80.2 COMPRESSOR SUCTION PRESSURE
          LIMITING, THROTTLING TYPE
          (COMPRESSOR SIDE)

    80.3 CONSTANT PRESSURE, SUCTION

    80.4 EVAPORATOR PRESSURE
          REGULATING, SNAP ACTION

    80.5 EVAPORATOR PRESSURE
          REGULATING, THERMOSTATIC
          THROTTLING TYPE

    80.6 EVAPORATOR PRESSURE
          REGULATING, THROTTLING TYPE
          (EVAPORATOR SIDE)

    80.7 HAND EXPANSION

    80.8 MAGNETIC STOP

    80.9 SNAP ACTION

80.10 SUCTION VAPOR REGULATING

80.11 THERMO SUCTION

80.12 THERMOSTATIC EXPANSION

80.13 WATER

81 VIBRATION ABSORBER, LINE

ANSI Graphic Symbols for Pipe Fittings, Valves and Piping
ASA Z32-2-3-1949 (R1953)

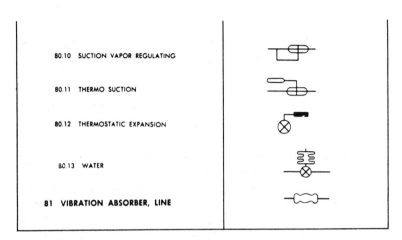

AMERICAN NATIONAL STANDARD					
	FLANGED	SCREWED	BELL & SPIGOT	WELDED	SOLDERED
1 BUSHING					
2 CAP					
3 CROSS					
3.1 REDUCING					
3.2 STRAIGHT SIZE					
4 CROSSOVER					
5 ELBOW					
5.1 45-DEGREE					

	FLANGED	SCREWED	BELL & SPIGOT	WELDED	SOLDERED
5.2 90-DEGREE					
5.3 TURNED DOWN					
5.4 TURNED UP					
5.5 BASE					
5.6 DOUBLE BRANCH					
5.7 LONG RADIUS					
5.8 REDUCING					
5.9 SIDE OUTLET (OUTLET DOWN)					
5.10 SIDE OUTLET (OUTLET UP)					
5.11 STREET					
6 JOINT					
6.1 CONNECTING PIPE					
6.2 EXPANSION					
7 LATERAL					
8 ORIFICE FLANGE					
9 REDUCING FLANGE					

	FLANGED	SCREWED	BELL & SPIGOT	WELDED	SOLDERED
10 PLUGS					
10.1 BULL PLUG					
10.2 PIPE PLUG					
11 REDUCER					
11.1 CONCENTRIC					
11.2 ECCENTRIC					
12 SLEEVE					
13 TEE					
13.1 (STRAIGHT SIZE)					
13.2 (OUTLET UP)					
13.3 (OUTLET DOWN)					
13.4 DOUBLE SWEEP)					
13.5 REDUCING					
13.6 SINGLE SWEEP)					
13.7 SIDE OUTLET (OUTLET DOWN)					
13.8 SIDE OUTLET (OUTLET UP)					
14 UNION					
15 ANGLE VALVE					
15.1 CHECK					

247

	FLANGED	SCREWED	BELL&SPIGOT	WELDED	SOLDERED
15.2 GATE ELEVATION					
15.3 GATE (PLAN)					
15.4. GLOBE (ELEVATION)					
15.5 GLOBE (PLAN)					
15.6 HOSE ANGLE	SAME AS	SYMBOL	23.1		
16 AUTOMATIC VALVE 16.1 BY-PASS					
16.2 GOVERNOR-OPERATED					
16.3 REDUCING					
17 CHECK VALVE					
17.1 ANGLE CHECK	SAME AS	SYMBOL	15.1		
17.2 (STRAIGHT WAY)					
18 COCK					
19 DIAPHRAGM VALVE					
20 FLOAT VALVE					
21 GATE VALVE					
*21.1					
21.2 ANGLE GATE	SAME AS	SYMBOLS	15.2 & 15.3		

	FLANGED	SCREWED	BELL&SPIGOT	WELDED	SOLDERED
21.3 HOSE GATE	SAME AS	SYMBOL	23.2		
21.4 MOTOR-OPERATED	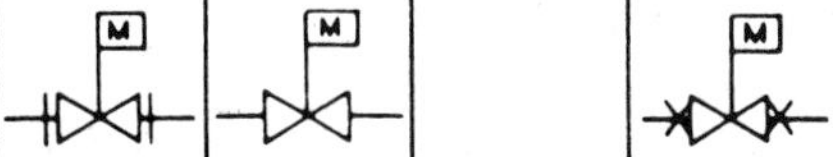				
22 GLOBE VALVE 22.1					
22.2 ANGLE GLOBE	SAME AS	SYMBOLS	15.4 & 15.5		
22.3 HOSE GLOBE	SAME AS	SYMBOL	23.3		
22.4 MOTOR-OPERATED	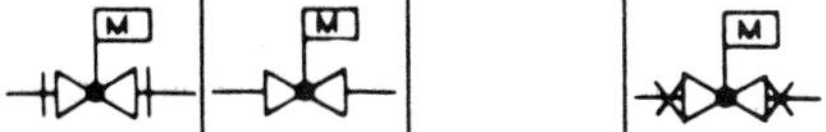				
23 HOSE VALVE					
23.1 ANGLE	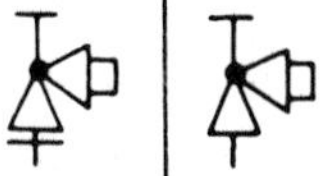				
23.2 GATE	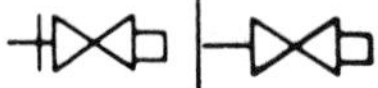				
23.3 GLOBE	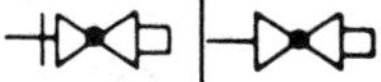				
24 LOCKSHIELD VALVE					
25 QUICK OPENING VALVE					
26 SAFETY VALVE					
27 STOP VALVE	SAME AS	SYMBOL	21.1		

AMERICAN NATIONAL STANDARD

AIR CONDITIONING

28 BRINE RETURN — — — BR — — —

29 BRINE SUPPLY ———— B ————

30 CIRCULATING CHILLED OR HOT WATER FLOW ———— CH ————

31 CIRCULATING CHILLED OR HOT-WATER RETURN— — —CHR— — —

32 CONDENSER WATER FLOW ———— C ————

33 CONDENSER WATER RETURN — — — CR — — —

34 DRAIN ——————— D ———————

35 HUMIDIFICATION LINE —— · —— H —— · ——

36 MAKE-UP WATER —— · —— · —— · ——

37 REFRIGERANT DISCHARGE ——————R D————————

38 REFRIGERANT LIQUID —————— RL————————

39 REFRIGERANT SUCTION — — —R S— — —

HEATING
40 AIR-RELIEF LINE — —— — — —— —

41 BOILER BLOW OFF —— —— —— ——

42 COMPRESSED AIR ——————— A ———————

43 CONDENSATE OR VACUUM PUMP DISCHARGE —o— —o— —o—

44 FEEDWATER PUMP DISCHARGE —oo— —oo— —oo—

45 FUEL-OIL FLOW ———————F O F————————

46 FUEL-OIL RETURN — — —F O R— — —

47 FUEL-OIL TANK VENT — — —F O V— — —

48 HIGH-PRESSURE RETURN — ⫫ — ⫫ — ⫫ —

49 HIGH-PRESSURE STEAM —⫫— ⫫ ⫫ —

50 HOT-WATER HEATING RETURN — — — — — ——

51 HOT-WATER HEATING SUPPLY ————————————

52 LOW-PRESSURE RETURN — — — — — — —

53 LOW-PRESSURE STEAM ————————————

54 MAKE-UP WATER ——· ——· ——· —

55 MEDIUM PRESSURE RETURN — ╪ — ╪ — ╪ —

56 MEDIUM PRESSURE STEAM —╪— ╪ ╪ —

PLUMBING
57 ACID WASTE ———— ACID ————

58 COLD WATER —·——·——·——·—

59 COMPRESSED AIR ——————— A ———————

60 DRINKING-WATER FLOW —— · —— · ——

61 DRINKING-WATER RETURN ——·· ——·· ——

62 FIRE LINE —F ——————— F ———

63 GAS —G ——————— G ———

64 HOT WATER ——·· ——·· —— ··—

65 HOT-WATER RETURN ——··· ——··· —— ···—

66 SOIL, WASTE OR LEADER (ABOVE GRADE) ————————————

67 SOIL, WASTE OR LEADER (BELOW GRADE)	— — — — — — —
68 VACUUM CLEANING	— V ——————— V ————
69 VENT	— — — — — — — —
PNEUMATIC TUBES 70 TUBE RUNS	══════════════
SPRINKLERS 71 BRANCH AND HEAD	——o———————o———
72 DRAIN	——s— — — —s ——
73 MAIN SUPPLIES	———————— s —————

Published by permission of the American Standards Graphical Symbols for Plumbing with the permission of the publisher, the American Society of Mechanical Engineers, United Engineering Center, 345 East 47th Street, New York, New York 10017. Portions extracted from the American National Standards Graphical Symbols for Pipe Fittings, Valves, and Piping (ASA Z32.2.3–1949) and from the American National Standards Graphical Symbols for Heating, Ventilating, and Air Conditioning (ASA Z32-2.4–1949).

251

CHAPTER 15

Architectural Drawings

Architectural drawings are based on the principles set forth in other parts of this book. They are a means of transferring the thoughts of the architect to the builders and any other craftsmen whose responsibility it is to construct the building. In the drawings, which are ordinarily called *plans,* there are various elevations shown, such as the front and side, as well as sectional and orthographic projections. Graphic symbols, as illustrated in chapter 14, are used to present what is needed and where the items are to be placed.

The different parts or drawings that are necessary to show the structure, such as the mechanical and electrical installations, are all shown graphically. Therefore, one must become familiar with these symbols, not only of one particular trade, but of the other trades as well. This is necessary so that complete coordination may be reached between the various trades. In the construction of a building, time and money may be saved by representatives of the different trades sitting down together and going over the plans and laying out the pattern to be followed. No trade can work independently of the other. If this is attempted, confusion results and some work will have to be redone to make all parts of the scheme fit together. During construction, the general contractor, the plumber, the steel workers, and the electrical and mechanical contractors must lay out the work together and decide who installs what, where, and when.

On the job, the architect usually has a representative present to assist in coordinating the work and in making the decisions that may be required. In the designing of the building, the owner or builder will often draw a rough sketch of his intentions, after which he sits down with the architect and they discuss what the owner will need and want in a building design. His requirements are noted as to space, machinery, electrical loads, numbers of persons that will occupy the building, and what the future requirements might be. The owner will sometimes provide a rough sketch of his ideas. These need not be drawn to scale or with any degree of accuracy; they are merely his ideas of what he might want. In this discussion, no elaborate plans are given. Simple plans are used, as they show the owner's intent and fall into line with more extensive plans.

Figure 15.1 illustrates the point of the owner's rough sketch of his ideas for the first floor. Figure 15.2 shows the ideas for the second floor. Note that there are no details shown, merely a sketch of an idea. After the sketch is drawn, the architect and owner can

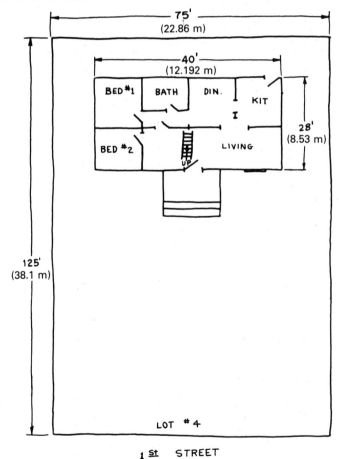

Fig. 15.1. An owner's sketch of the first floor of a residence.

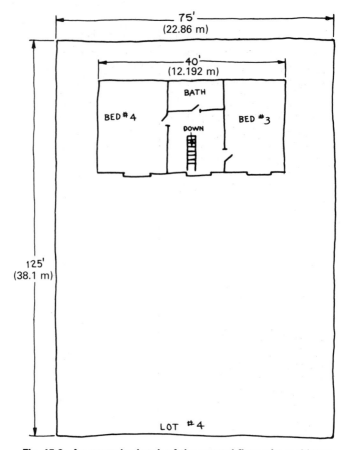

Fig. 15.2. An owner's sketch of the second floor of a residence.

sit down and discuss details, at which time no doubt another, more detailed free-hand sketch will be drawn. When a tentative solution is reached, the architect will draw a sketch as shown in figure 15.3. In addition, he will have to take into consideration elevation, sewers, water lines, and gas lines. After a tentative decision is reached, the architect may be asked to sketch an outline showing how the structure will look when completed.

When the owner has signed construction papers, the architect will begin to prepare the final plans. There are many preliminary things to do, such as surveying the land to see how much excavating will be required. The location of the property lines and the general drainage plan for the immediate vicinity must be considered, and

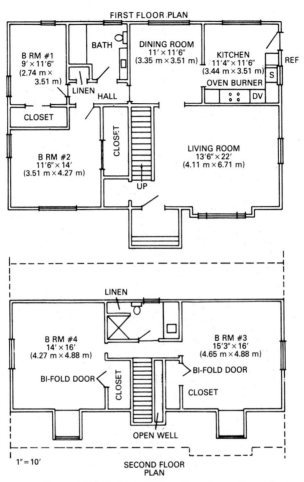

Fig. 15.3. An architect's drawing (preliminary), without dimensions, of the owner's idea for the residence shown in figs. 15.1. and 15.2.

the water, sewer, gas, and telephone and power lines that exist must also be considered. Local regulations regarding types of construction permitted, set backs, etc., must all be taken into consideration.

The final drawings are sent to the plan checkers of the inspection departments that have local jurisdiction. Here they are checked to see that they conform to local codes. Corrections are noted or the plans are approved. Most specifications that accompany plans put the burden of following local codes that are applicable on the

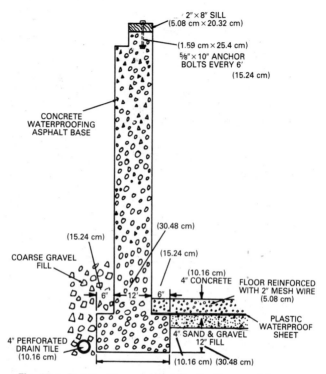

Fig. 15.4. Detail of basement walls, footings, and floor.

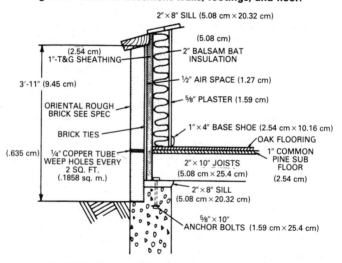

Fig. 15.5. Details of floor, brick veneering, etc.

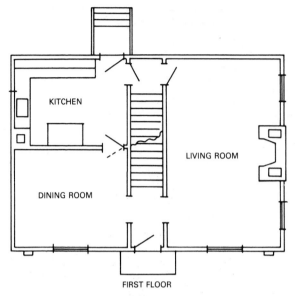

FIRST FLOOR

Fig. 15.6. A typical first-floor plan.

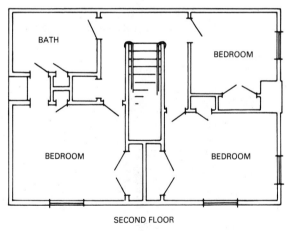

SECOND FLOOR

Fig. 15.7. A typical second-floor plan.

tradesmen and contractors. When questions arise, such as an electrical or mechanical problem, the contractor involved takes these problems up with the architect or his assistant who, in turn, takes them to the engineer that has performed the design work.

There are various methods of bidding on plans. Sometimes,

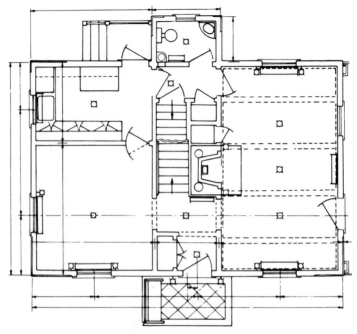

Fig. 15.8. A typical first-floor plan with dimensions.

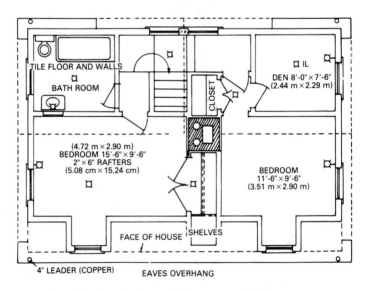

Fig. 15.9. A typical second-floor plan with dimensions.

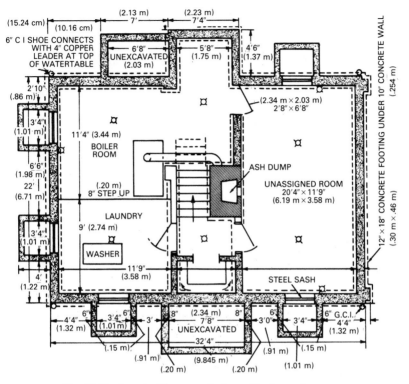

Fig. 15.10. A typical basement plan with dimensions.

the general contractor gives the entire bid, but calls for bids from subcontractors. This method has some advantages in that the general contractor is responsible for the entire job.

At times, the general contractor and each subcontractor bid their parts separately. This method has an advantage; the general contractor's percentage is removed from the subcontractors' bidding, and the owner has more control over who gets the bids. The details of the architect's sketches for the preliminary part are not important to the mechanic; therefore, they will not be covered in this book. The mechanic is mainly interested in how to perform his part of the work. In this chapter, most of our efforts have been on details covering construction, and most of the drawings shown have followed this type of pattern.

Figure 15.4 illustrates a typical detail of the basement wall, footings, and floor. Figure 15.5 shows a typical detail of the floor joists, brick veneering, etc., as it attaches to the foundation shown in figure 15.4. A typical floor plan of the first floor and second floor is shown in figures 15.6–7. Figures 15.8–9 show a typical floor plan of a first and second floor, with dimensions added. Figure 15.10 is a typical plan of a basement, with dimensions added.

CHAPTER 16

Architects' Conventions

Drawings, whether for machinists or builders, are essentially similar; the main difference is in the symbols. With architectural drawings, especially on larger jobs, there is a set of specifications, commonly known as *spec sheets*. These specifications incorporate details that are not readily shown on the prints. There are codes to be followed, bonds to be posted, determination of who is responsible for the work, complete details on equipment to be furnished, and any other details that would clutter up the drawings. The *spec sheet* becomes a part of the job as much as the drawings or blueprints.

Smaller jobs do not go into so much detail on specifications. These specifications are divided into structural, mechanical, electrical, or any other parts that may be required. Any changes made are listed on an *addendum,* which becomes a part of the specifications. All changes from the original drawings and specifications should have signatures on the drawings and written orders for such changes. This saves much confusion when the settlement date comes. A thorough understanding of the procedures is necessary in order that the tradesmen may protect the interest of their employers.

Frequently, drawings are not as completely dimensioned as they should be and have to be scaled by the mechanic. This is not the best practice, as the exact size of the prints may vary and the locations of objects may not come out correctly. Should a drawing be vague in dimensions, the architect should be consulted for the proper dimensions. As covered in chapter 14, symbols are used as a shorthand for the simplification of drawings. These symbols are universal and the tradesman should become thoroughly familiar with those used in his field of work. Architectural symbols or conventions include the following:

1. Building details
2. Plumbing (covered in chapter 14)
3. Piping (covered in chapter 14)
4. Welding
5. Electrical (covered in chapter 14)

BUILDING CONVENTIONS

Walls of frame buildings are represented on floor plans by two parallel lines spaced at a distance apart equal to the wall thickness, as "A" in figure 16.1. Masonry walls are shown on a floor plan by cross sectional lines, as shown in "B" (figure 16.1). Walls of all types of construction may also be shown as in "C" (figure 16.1), that is, by heavy dark lines which save time in drawing and give a better print.

There are many and varied types of window construction, the symbols for some being shown in "D" through "J" (figure 16.1). The specifications should show the materials and types of construction, thickness of glass, and type of glass to be used. These details may be listed as a supplement to the specifications, or on the drawings if there is room.

There are many types of doors; the symbols for some of these are shown in "K" through "Q" (figure 16.1). Where special doors are called for, a drawing showing the detail should accompany the main drawing. Detailed sketches or drawing inserts should show all details of sills, especially where masonry construction is to be used. In drawings, the dashed line should be avoided where it is intended to indicate some part that is in view; the dashed line is ordinarily intended to represent some hidden feature or part.

A few conventions or symbols for chimneys and fireplaces are shown in figure 16.2. There may be special features which should be shown in additional drawings. In each instance where details are required, a notation should be added referring to the detail drawings.

Stairs must be identified as to their direction, and whether they are cased or open. Some methods of identification are shown in figure 16.3. Arrows show the direction of the stairs.

Symbols for the identification of materials are shown in figure 16.4. Structural steel work in a building is often calculated by engineers. The standards for steel are established by the American

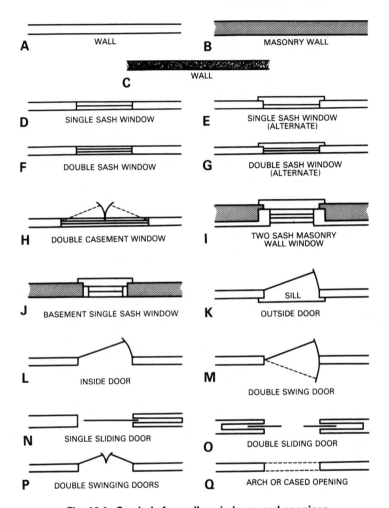

A WALL

B MASONRY WALL

C WALL

D SINGLE SASH WINDOW

E SINGLE SASH WINDOW (ALTERNATE)

F DOUBLE SASH WINDOW

G DOUBLE SASH WINDOW (ALTERNATE)

H DOUBLE CASEMENT WINDOW

I TWO SASH MASONRY WALL WINDOW

J BASEMENT SINGLE SASH WINDOW

K OUTSIDE DOOR / SILL

L INSIDE DOOR

M DOUBLE SWING DOOR

N SINGLE SLIDING DOOR

O DOUBLE SLIDING DOOR

P DOUBLE SWINGING DOORS

Q ARCH OR CASED OPENING

Fig. 16.1. Symbols for walls, windows, and openings.

Institute of Steel Construction, Inc., 101 Park Avenue, New York, New York, and all of the specifications are printed in a handbook. In the drawings, reference will be made to the AISC specifications. The actual figuring is an engineering study in itself. Angles of steel are given by the bevel, which gives the number of inches in one direction that it slopes for 12 inches in a perpendicular direction, as shown in figure 16.5. Standard steel shapes generally used are the standard I beam, wide flange beam, channel, angle, tee, and zee bar, (see fig. 16.6). I beams, beams, and channels are indicated

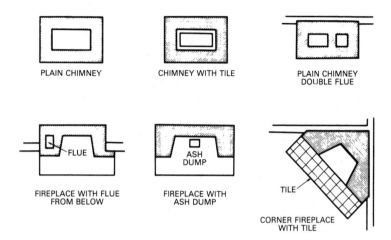

Fig. 16.2. Symbols for chimneys and fireplaces.

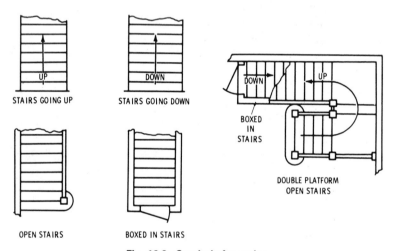

Fig. 16.3. Symbols for stairs.

by giving the depth and the weight per foot. When the steel member is given, it may be identified from the AISC handbook; however, many abbreviations will often be given as follows:

Center line (cl)   Out to out (o to o)

Center to center (c to c)   Diameter (ϕ)

On center (oc)   Pounds (#)

264

Back to back (b to b)

Wide flange (wf)

Front (fr)

Between center (bc)

Center (ctr)

Outstanding leg of angle (osl)

Symmetrical about (symm abt)

Arrangement (arrgt)

Architecture (arch)

Weight (wt)

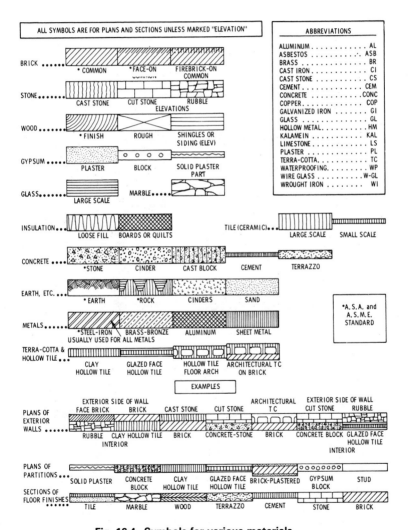

Fig. 16.4. Symbols for various materials.

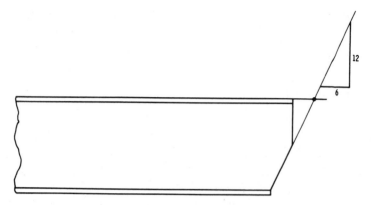

Fig. 16.5. A method of indicating an angle on standard steel drawings.

STRUCTURAL STEEL DRAWINGS

There are two classes of structural steel drawings: (1) general design drawings indicating, by various plans, elevations and sections, the general arrangement of the structure, and the design of the various

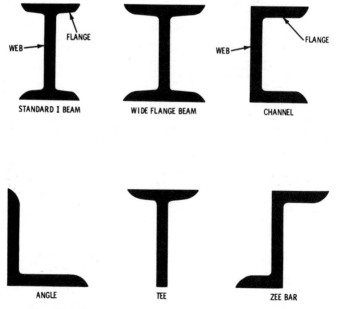

Fig. 16.6. Representation of standard structural shapes.

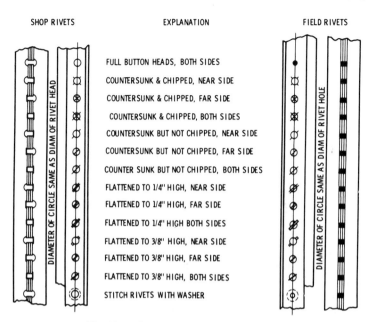

Fig. 16.7. Shop and field rivet symbols.

members making up the structure; and (2) shop drawings that are made in the drafting room of the shop where the steel members are fabricated.

It would be impossible to include in this book all of the dimensions, configurations, weights, and strengths, of the structural steel that is used in building. This information will be found in the Manual of Steel Construction published by the American institute of Steel Construction, Inc. A few examples will be included here to help you in the interpretation of some of the information that will be found on blueprints.

Designations of Structural Steel

American Standard Beams	15 I 42.9 #
Wide Flange Sections	24 wf 74 #·
American Standard Channels	9 ☐ 13.4 #
Angles	L 6 × 4 × ½

Tees	T $3 \times 3 \times 6.7$ #
Zees	Z $6 \times 3\frac{1}{2} \times 15.7$
Plates	P $16 \times \frac{1}{2}$
Bars, Square	BAR $1\frac{1}{4}$ □
Bars, Round	BAR 1 ○
Bars, Flat	BAR $3 \times \frac{1}{4}$

Represents pounds

FABRICATING A STRUCTURE

To make up a structure, the various shapes are put together by bolting, welding, or riveting. Since structural floor plans are drawn on a small scale, a beam or girder is represented by a single thick line. Rivets may be driven in at the shop or in the field, and are shown on blueprints, as in figure 16.7.

Reinforcing bar for concrete footings and floors are listed by numbers instead of by diameter. These are cast items, with raised portions for holding in the concrete. These reinforcing bars were formerly designated by inches; they are now designated by numbers. Dimensions are approximate.

Table 16.2. Dimensions for Concrete Reinforcing Bars.

Deformed designation bar number	Unit lbs. per ft. Weight	Nominal dimensions round sections		
		Diameter, ins.	Cross-Sectional Area Sq. Ins.	Perimeter, ins.
3	0.376	0.375	0.11	1.178
4	0.668	0.500	0.20	1.571
5	1.043	0.625	0.31	1.963
6	1.502	0.750	0.44	2.356
7	2.044	0.875	0.60	2.749
8	2.670	1.000	0.79	3.142
9	3.400	1.128	1.00	3.544
10	4.303	1.270	1.27	3.990
11	5.313	1.410	1.56	4.430

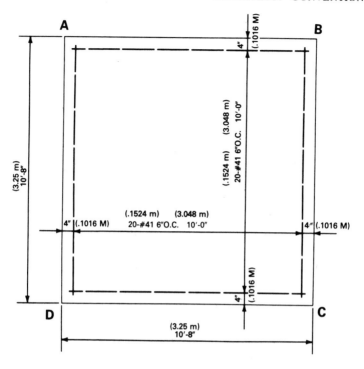

Fig. 16.8. Illustrating symbols used for reinforced concrete footings.

REINFORCED CONCRETE SYMBOLS AND ABBREVIATIONS

What do the heavy dashed lines represent?

Answer: They represent reinforcing rods in accordance with the usual system of representation. The 20 means that there are 20 of these rods to be placed; the two end ones are shown here. The other 18 are to go in between these and to be spaced 6 inches apart in accordance with the notation 6″ oc (on center). The 1/2″ φ means that the rods are to be 1/2″ in diameter. If they were to be 3/4″ square rods they would be indicated as 3/4″☐. On some drawings of reinforced concrete structures, the rods are shown as heavy solid lines rather than dashed lines—a timesaving idea.

Plates	P $16 \times \frac{1}{2}$
Bars, Square	BAR $1\frac{1}{4}$ □
Bars, Round	BAR 1 ○
Bars, Flat	BAR $3 \times \frac{1}{4}$

CHAPTER 17

Electrical Blueprints

A list of graphic symbols for electrical blueprints appears in chapter 14 and should be used with this chapter. A few generalized sketches will be shown here to provide familiarity with the subject. Figures 17.1–2 show plans for a residence. Figure 17.1 is the main floor; Figure 17.2 is the basement plan. The following references will be tied to these two illustrations.

Service Entrance:

2″ rigid conduit mast, with 3 No. 0 copper conductors, THW type panel.

36-circuit minimum, with 150-ampere buses and a 150-ampere main circuit breaker.

Circuit No. 1, supplying receptacles in bedroom No. 1 and lights in bedroom No. 2. sp 15-ampere breaker, No. 12 NM cable.

Circuit No. 2, supplying receptacles in bedroom No. 2 and lighting in bedroom No. 1. sp 15-ampere breaker, No. 12 NM cable.[1]

Circuit No. 3, 1500-watt, 240-volt bathroom heater, dc 15-ampere breaker.

Circuit No. 4, lights and receptacles in bath and hall, and lights in bedroom No. 3. sp 15-ampere breaker, No. 12 NM cable.

Circuit No. 5, dp 40-ampere breaker and No. 8 SE cable to supply dryer.

1. On 15-ampere circuits, No. 14 copper NM cable would be sufficient, but No. 12 is specified for good voltage regulation.

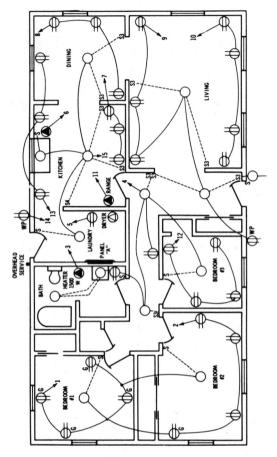

Fig. 17.1. An electrical hookup circuit of a one-story residence.

Circuit No. 6, dp 20-ampere, multiwire circuit to supply built-in dishwasher and garbage disposal, No. 12-3 wire NC cable.[2]

Circuit No. 7, sp 20-ampere breaker, No. 12 NM cable for appliance receptacles in kitchen.

Circuit No. 8, sp 20-ampere breaker, No. 12 NM cable for diningroom receptacles.

Circuit No. 9, sp 20-ampere breaker, No. 12 NM cable, for

2. All NM cable used is to be with ground.

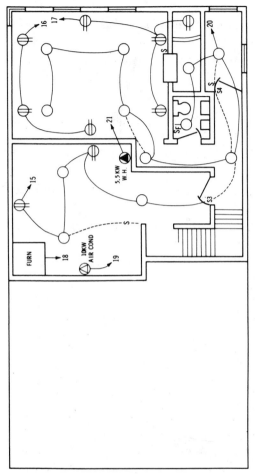

Fig. 17.2. An electrical hookup of a basement.

WP receptacle at front of house, receptacle in living room, hall light, and living-room light.

Circuit No. 10, sp 20-ampere breaker, No. 12 NM cable for receptacles in living room.

Circuit No. 11, dp 50-ampere breaker, No. 6-3 SE cable for range.

Circuit No. 12, receptacles in bedroom No. 3, lights in laundry, dining room and kitchen, sp 15-ampere breaker, No. 12 NM cable.

ANSWERS ON BLUEPRINT READING

Circuit No. 13, sp 20-ampere breaker, No. 12 NM cable for appliance receptacle in kitchen and dining room.

Circuit No. 14, sp 20-ampere breaker, No. 12 NM cable for wp receptacle in back yard and future yard lights.

Circuit No. 15, sp 15-ampere breaker, No. 14 NM cable, for furnace lights, receptacles and basement hall light.

Circuit No. 16, sp 20-ampere breaker, No. 12 NM cable for receptacles in family room.

Circuit No. 17, 20-ampere breaker sp, No. 12 NM cable for receptacles in family room.

Circuit No. 18, sp 15-ampere breaker for furnace, No. 12 NM cable.

Circuit No. 19, dp 70-ampere breaker, No. 4 NM cable with ground for air conditioner.

Circuit No. 20, sp 15-ampere breaker, No. 12 NM cable for halls in basement.

Circuit No. 21, dp 30-ampere breaker, No. 10 NM cable with ground for water heater.

There are 6 dp (double pole) breakers and 19 sp (single pole) breakers that would use 31 spaces in the electrical cabinet; and a minimum of 32 are required. The next size larger electrical box should therefore be used so the circuits will not be overloaded; extra circuits will then be available for future expansion.

Unless specified, 12 inches is a good height for most receptacles, except in the kitchen. A good height for switches is 46 inches. The National Electrical Code must be followed throughout the installation of the electrical system, unless local regulations are more stringent. Figure 17.3 shows a one-line diagram of a power and lighting layout. One-line diagrams are used for the purpose of showing, at a glance, the intent of the layout. It is not used for the actual layout, but is helpful in showing the engineer's intentions. In this particular case, there is a 1,200 to 2,400/4,160 wye-wye pad-mount transformer located outside of the building; nothing is specified as to the metering.

Underground service uses a 4-wire wye that goes to the primary switch gear in the building. An overload current protector supplies 2,400/4,160 volts to a 120/208-volt dry transformer, supplying panel No. 2 which has four sets of overload current protectors supplying

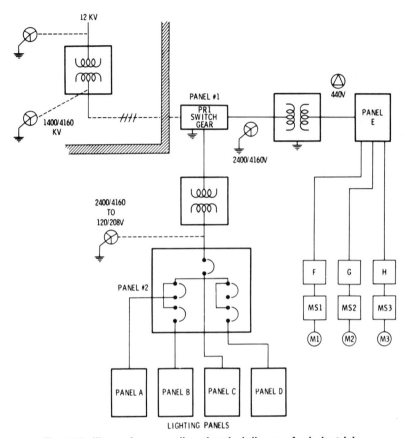

Fig. 17.3. Illustrating a one-line electrical diagram for industrial use.

lighting panels *A*, *B*, *C*, and *D*. There is nothing mentioned as to the specifications of the panels, wire, conduit, etc. This will be detailed in the other prints and specifications. One branch from panel No. 1 feeds a 2,400/4,160-volt to 440-volt, wye-delta, dry transformer, with the 440-volt side ungrounded. This, in turn, feeds panel *E*, which will have a main breaker and three or more branch circuits supplying panel *F*. Panel *F* is a motor branch circuit overload current device, with magnetic starter (MS 1) supplying motor M1.

Two other motors, M2 and M3, are supplied in the same manner. Specifications will be either on the detail print, on a schedule, or in the *specs* for the job. At a glance, a great deal is to be learned from one-line diagrams providing not too much is taken for granted.

275

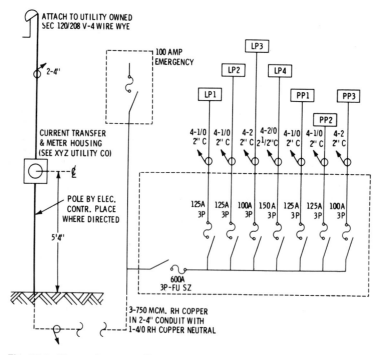

Fig. 17.4. Illustrating a one-line electrical diagram for residential housing.

Figure 17.4 illustrates another one-line diagram that shows a little more detail. Starting at the service entrance, the system is a 120/208-volt wye, two runs of 4-inch conduit, each with 3-750 MCM phase conductors type RH, with a 4/0 neutral, all of copper, with part of the service underground. The current transformers and meter are connected according to the utility company's specs, with the meter located 5 foot 4 inches to centerline from grade. This supplies a 100-ampere emergency service panel ahead of a 600-ampere, 3-pole, solid-neutral, fused disconnect. This main panel has seven feeder circuits contained in the same panel. They consist of the following:

1. A 3-pole, solid-neutral, 125-ampere, fused switch, feeding 2″ conduit, with 4 No. 1/0 copper conductors to supply panel LP1.

2. A 3-pole, solid-neutral, 125-ampere, fused switch, feeding 2″ conduit, and 4 No. 1/0 copper conductors supplying feeder panel LP2.

3. A 3-pole, solid-neutral, 100-ampere, fused switch, feeding 2″ conduit, and 4 No. 2 copper conductors to supply feeder panel LP3.

4. A 3-pole, solid-neutral, 150-ampere, fused switch, feeding 2½″ conduit, and 4 No. 2/0 copper conductors to supply feeder panel LP4.

5. A 3-pole, solid-neutral, 125-ampere, fused switch, feeding 2″ conduit, and 4 No. 2 copper conductors to supply feeder panel PP1.

6. A 3-pole, solid-neutral, 125-ampere, fused switch, feeding 2″ conduit, and 4 No. 1/0 copper conductors to supply feeder panel PP2.

7. A 3-pole, solid-neutral, 100-ampere, fused switch, feeding 2″ conduit, and 4 No. 1/0 copper conductors to supply feeder panel PP3.

LP indicates *lighting panels*; PP indicates *power panels*. The power panels will have no lighting from them. Another one-line diagram is shown in figure 17.5, to illustrate the telephone conduit. The telephone service comes into the building in a 1½″ conduit buried 30″ deep. This terminates in box T-1, from which the various

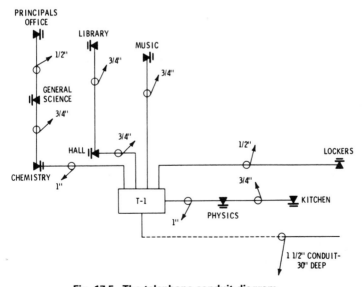

Fig. 17.5. The telephone conduit diagram.

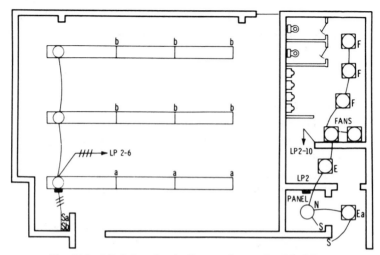

Fig. 17.6. A lighting circuit diagram for a school building.

circuits branch out. A 1″ conduit leads to the chemistry room for a phone, and from there it reduces to ¾″ conduit to the general science room, where it reduces to ½″ conduit to supply the principal's office. The next conduit from T-1 is the ¾″ conduit that leads to a hall phone, and from there with ¾″ conduit to the library. The next is

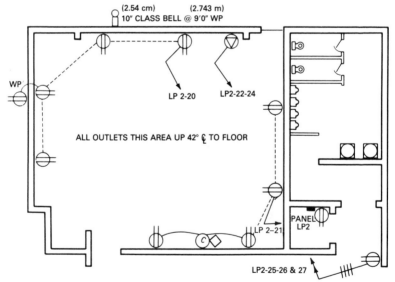

Fig. 17.7. Receptacle outlets illustrated for a school building.

a ¾″ conduit that leads to the music room. The fourth conduit leads to the locker room and is ½″ in diameter. The fifth conduit is 1″ and goes to the physics room, where it reduces to ¾″ and goes to the kitchen.

Figure 17.6 illustrates the lighting plans for one classroom and the boys' rest room in a school building. In the classroom, there are 9 fluorescent fixtures, indicated as *a* and *b*, the types of fixtures to be specified on the fixture schedule. These fixtures are supplied

CKT	POLE	BRKR	LOAD	PHASE	LOAD	BRKR	POLE	CKT
				LP 2				
1	1	20	1620	A	1620	20	1	2
3	1	20	1200	B	650	20	1	4
5	1	20	675	C	1620	20	1	6
7	1	20	1250	A	SPARE	20	1	8
9	1	20	1620	B	875	20	1	10
11	1	20	1620	C	540	20	1	12
13	1	20	840	A	1440	20	1	14
15	1	20	1440	B	1440	20	1	16
17	1	20	1440	C	900	20	1	18
19	1	20	720	A	900	20	1	20
21	1	20	720	B	KILN	30	2	22
23	1	20	900	C				24
25	2	20	UTILITY	A	UTILITY			26
27				B				28
29	2	20	1/2 HP	C	3 HP	20	3	30
31				A				32
33	1	20	SPARE	B	3 HP	20	3	34
35	1	20	SPARE	C				36
37	1	20	SPARE	A				38
39			SPACE ONLY					40
41			SPACE ONLY					42

CONNECTED LOAD : 36 KW
DESIGN LOAD : 40 KW
FEED : 4 No. 1/0 TWIN 2″ COND. 125 AMP SERVICE
PANEL : 42 CKT. 225 AMP MAIN LUGS ONLY SURFACE MOUNTED 3d 4 WIRE. N. QO, 42-4L

Fig. 17.8. Illustrating the typical wiring table for a 42-circuit panel.

ANSWERS ON BLUEPRINT READING

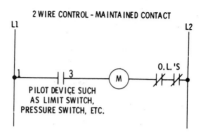

Fig. 17.9. A diagram of a magnetic starter with two overload units.

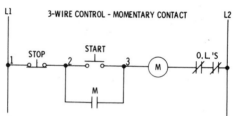

Fig. 17.10. A diagram of a magnetic starter with a pushbutton start and stop station.

from panel LP2 on circuit 6, by means of four conductors—two phase wires and two neutrals. The junction is made in the first row of fixtures that are marked *a;* from here, the switch legs go to S*a* and S*b*. S*a* is supplying voltage to fixtures *a,* and S*b* is supplying voltage to fixtures *b.* There are two switch conductors and two hot conductors supplying the switches. In addition, mention is made of 220 volts for an electric kiln.

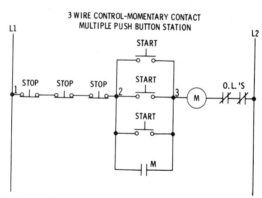

Fig. 17.11. Illustrating a magnetic starter with three push-button start-and-stop station.

280

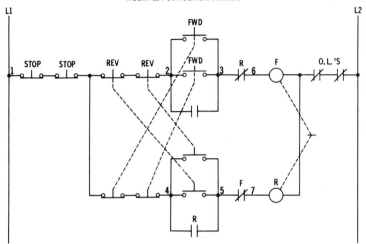

**3 WIRE CONTROL - REVERSING STARTER
MULTIPLE PUSH BUTTON STATION**

MORE THAN ONE "FORWARD- REVERSE-STOP" PUSH BUTTON STATION
MAY BE REQUIRED AND CAN BE CONNECTED IN THE MANNER SHOWN ABOVE.

Fig. 17.12. Illustrating a magnetic reversing starter with a multiple control.

In the boys' rest room, the fixtures F, E, E_a, and n will be found on a fixture schedule. Note that these lights, and the two exhaust fans, are fed from panel LP2 on circuit 10. Note also the location of panel LP2. When the fixtures F are turned on, the fans also come on. E, N, and E_a are switched independently.

Note the lack of outlets on this print. The engineer will draw

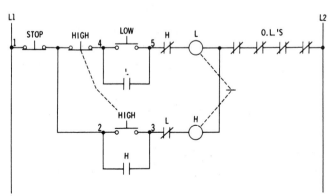

Fig. 17.13. Illustrating a 3-wire control circuit for a magnetic starter with a two-speed motor.

another layout for these so as not to complicate the drawings. Figure 17.7 shows the same area as in figure 17.6, but with the outlets included. All outlet receptacles are 42″ from the floor. The electric kiln is fed by circuits 22 and 24 from panel LP2. Five receptacles (one being outdoors and weatherproof) are fed from circuit 20 of panel LP2. A clock and four receptacles are fed from circuit 21 of panel LP2. One receptacle is being fed from circuits 25, 26, and 27 of panel LP2. This would mean that three phase wires and a neutral are run to this outlet, and then from there they feed other parts of the building.

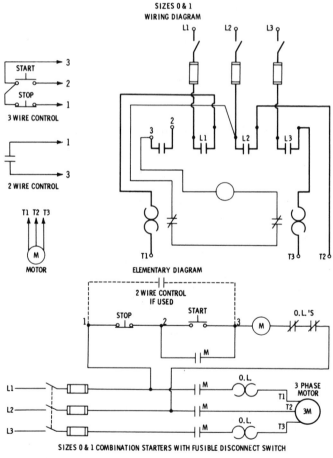

Fig. 17.14. Illustrating a wiring diagram for a magnetic starter with a push button.

Figure 17.8 illustrates the schedule for panel LP2. This is a 42-circuit panel, which is the limit that is permitted by the National Electrical Code for panels that service lighting and appliances. Each circuit (ckt) is numbered, which includes the number of poles, the size of the breaker in amperes, the load on the circuit, and the phase that it connects to, such as the *A*, *B*, and *C* phases of the 4-wire wye system. Figure 17.4 shows that this panel is fed from a 125-ampere feeder circuit. This illustration will assist in the understanding of the purpose of the one-line diagrams. The type of building will determine the actual running of the conduits or cables.

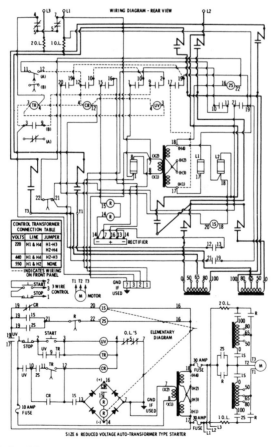

Fig. 17.15. A wiring diagram of a reduced-voltage autotransformer type of starter.

There are problems that must be worked out on the job in cooperation with all the tradesmen involved as well as the architect.

The next few illustrations are of motor starters. Figure 17.9 illustrates the diagram of a magnetic starter with two overload units and a pilot device, such as a pressure switch. Figure 17.10 illustrates a magnetic starter with a pushbutton start-and-stop station, two overloads, and holding contacts. Figure 17.11 shows a magnetic starter like the one in figure 17.10 except that there are three pushbutton start-and-stop stations. Figure 17.12 illustrates a 3-wire control reversing magnetic starter with multiple push-button stations. Figure 17.13 shows a 3-wire control, magnetic starter for two-speed operation of a motor. Figure 17.14 illustrates two types of diagrams—a *wiring diagram* and an *elementary* or *schematic diagram*—for a magnetic starter with one pushbutton station. This is shown to illustrate the two different types of diagrams that will be found in the field. Two types of diagrams for a *reduced-voltage autotransformer* type of starter are shown in figure 17.15.

CHAPTER 18

The Slide Rule Calculator

It might at first seem out of place to include the slide rule in a book pertaining to blueprints, but this most certainly is not the case. The slide rule is indispensible for anyone connected with the building trades. It is an instrument that frightens many individuals; the slide rule calculator, however, is easy to learn to use (see figure 18.1).

Modern electronics has made a slide rule calculator available to us that will do anything the slide rule will do and that has many more functions than the slide rule. One might say that the slide rule is obsolete. With the calculator, arriving at an answer does not involve the eye's accuracy. The answer on the pocket slide rule calculator removes all of the guesswork associated with the use of the slide rule. There are many calculators on the market, and they are designed to do various types of calculations. Slide rule calculators that have replaced the slide rule are slightly higher in price than the calculators intended primarily for the basic functions of addition, subtraction, multiplication. and division.

The slide rule calculator is valuable to any person who works with figures and calculations. It is an instrument not only for the engineer but also for those individuals who have to make a number of complicated calculations involved with their work.

How does the slide rule calculator work?

Answer: The slide rule calculator illustrated in figure 18.1 operates on a system that allows a simple, straightforward entry of most problems. The system, called the Algebraic Operating System, is a development of Texas Instruments. There are 50 functions on this calculator, (see fig. 18.2).

How is the slide rule calculator powered?

Answer: Some are powered with batteries and some are powered with solar power cells.

285

Fig. 18.1. Slide rule calculator. *Courtesy Texas Instruments*

How does the calculator operate for data entry?

Answer: For maximum versatility, the calculator operates with a floating decimal point. It has digit keys to enter numbers from 0 through 9. It has *a decimal point key* to enter a decimal point. In addition, it has a *change sign key*, which when pressed after a number entry or a calculation changes the sign of the displayed number. A *pi key* is also provided so the value of pi can be entered correct to nine digits; for the display, however, the number is rounded to eight digits (3.1415927).

What are the basic arithmetic functions that can be performed on the slide rule calculator?

Answer: Addition, subtraction, multiplication, and division. These functions are completed with the basic keys illustrated in figure 18.3.

What can be done to correct an error that was put into the calculator?

Answer: At any point in a calculation, an incorrect number entry can be cleared without affecting any calculation that is in progress. An incorrect operation that has been entered into the

- 50 Calculator Functions

Arithmetic	$+, -, \times, \div$	4
Data Entry	$+/-, \pi$	2
Display	Scientific notation	1
Algebraic	$x^2, \sqrt{x}, 1/x, y^x, \sqrt[x]{y}, x!$	6
Clearing	Clear, Clear Entry, All Clear	3
Data Grouping	AOS algebraic operating system. Open and close parentheses (up to 15), and full algebraic hierarchy (up to 4 pending operations).	3
Memory	One memory with store, recall, sum, and exchange	4
Percent	$\%, +\%, -\%, x\%, \div\%$	5
Trigonometric	Sin, Cos, Tan, $\text{Sin}^{-1}, \text{Cos}^{-1}, \text{Tan}^{-1}$, and 3 angular modes (Degrees, Radians, Grads)	9
Angular Conversion	Degrees to Radians to Grads	3
Logarithmic	lnx, log, e^x, 10^x	4
Constant	Operates with $+, -,$ $\times, \div, y^x,$ and $\sqrt[x]{y}$	6
		50

- Accuracy—The internal calculating capacity is 9 digits even though only 8 can be displayed. The 8-digit displayed number is generally rounded to within ± 1 in the 8th digit for all functions except where noted.

Fig. 18.2. Fifty calculator functions that can be performed on the slide rule calculator. *Courtesy Texas Instruments*

calculator also can be completely cleared, allowing the operator to restart the problem.

How can operations be combined on the calculator?

Answer: Operations can be combined on the calculator by using the results from one calculation and then combining them with a second calculation to complete the calculations involved in the problem (see fig. 18.4).

Basic Keys

⊞ Add Key—Completes any previously entered +, −, ×, ÷, y^x, or $x\sqrt{y}$ function when not separated by an open parenthesis and instructs the calculator to add the next entered quantity to the displayed number.

⊟ Subtract key—Completes any previously entered +, −, ×, ÷, y^x, or $x\sqrt{y}$ function when not separated by an open parenthesis and instructs the calculator to subtract the next entered quantity from the displayed number.

⊠ Multiply Key—Completes any previously entered ×, ÷, y^x or $x\sqrt{y}$ function when not separated by an open parenthesis and instructs the calculator to multiply the displayed number by the next entered quantity.

⊞ Divide Key—Completes any previously entered ×, ÷, y^x or $x\sqrt{y}$ function when not separated by an open parenthesis and instructs the calculator to divide the displayed number by the next entered quantity.

⊟ Equals Key—Combines all previously entered numbers and operations. This key is used to obtain both intermediate and final results.

Fig. 18.3. Basic keys for arithmetic functions. *Courtesy Texas Instruments*

What is a calculator hierarchy?

Answer: A calculator hierarchy provides the calculator with a fixed set of standard algebraic rules. These rules are used to assign priorities to the various mathematical operations that have to be performed. If there were not a set of rules, mathematical expressions could have a number of different meanings. Figure 18.5 illustrates the interpretative order using the fixed set of standard algebraic rules.

How are the *parentheses keys* used in the mathematical calculations?

Answer: These keys are used to isolate particular numerical expressions for separate mathematical interpretation. There are calculations that require this. If this were not done, a large number of independent steps would be required to perform the calculation. Figure 18.6 illustrates the effective use of the parenthese keys in solving this problem.

Combining Operations

After a result is obtained in one calculation, it may be directly used as the first number in a second calculation. There is no need to reenter the number from the keyboard.

Example: 1.84 + 0.39 = 2.23 then
(1.84 + 0.39) ÷ 365 = 0.0061096

Enter	Press	Display	Comments
1.84	➕	1.84	
.39	➡	2.23	1.84 + 0.39
	➕	2.23	
365	➡	0.0061096	2.23 ÷ 365

Fig. 18.4. Procedure for combining functions. *Courtesy Texas Instruments*

How are calculations with a constant performed?

Answer: The *constant key* is used to store a number and its associated operation for repetitive calculations. With this feature, repetitive calculations can be simplified through the use of the constant feature of the calculator.

What are some of the special functions of the slide rule calculator?

Answer: The slide rule calculator has a number of special functions that can be used in a number of different types of calculations that are necessary in the types of activities covered in this book. For example, in calculating roots and powers, there is a *square key* and a *square root key* (see figure 18.7). "In addition, there is a y to the xth *power key* that raises the displayed value y to the xth power. A number of other special functions are discussed in this chapter.

What is the purpose of the reciprocal key?

Answer: The *reciprocal key* enables the user to divide the displayed value x into 1. For example enter 5 and press the *reciprocal key* $1/x$ and .2 will be displayed. Depress the key again, 5 will appear in the display.

What is the purpose of the factorial key?

Answer: The *factorial key* calculates the factorial(x) $(x-1)(x-2)...(2)(1)$ of the value x in the display for integers $0<x>69.01=1$ by definition. For example, $36! = 3.7199 \times 10^{41}$. When 36 is entered, the inv and $x!$ keys are depressed and 3.7199 41 will appear in the display.

Example: $4 \times (5 + 9) \div (7 - 4)(2 + 3) = 0.2304527$

Key in this expression and follow the path to completion.

Enter	Press	Display	Comments
4	[×][( ]	4.	(4 ×) stored pending evaluation of parentheses.
5	[+]	5.	(5 +) stored.
9	[ )]	14.	(5 + 9) evaluated.
	[+]	56.	Hierarchy evaluates (4 × 14).
	[( ]	56.	(56 ÷) stored pending evaluation of parentheses.
7	[−]	7.	(7 −) stored.
4	[ )]	3.	(7 − 4) evaluated.
	[yˣ][( ]	3.	Prepares for exponent.
2	[+]	2.	
3	[ )]	5.	(2 + 3) evaluated.
	[=]	0.2304527	(7 − 4)(2 + 3) evaluated; then divided into 4 × (5 + 9).

Fig. 18.5. An example of a complex calculation that can be performed on the slide rule calculator. *Courtesy Texas Instruments*

What is the purpose of the percent key?

Answer: The *percent key* is used to convert the number displayed from a percentage to a decimal. This function has a number of practical applications in blueprint work.

What are some other special functions that are usually found on the more advanced slide rule calculators?

Answer: Many of the advanced slide rule calculators can be used to calculate natural logarithms and natural antiloga-

Example: $5 + (8 \div (9 - (2 \div 3))) = 5.96$

Enter	Press	Display	Comments
5	(+)(()	5.	(5 +) stored
8	(÷)(()	8.	(8 ÷) stored
9	(−)(()	9.	(9 −) stored
2	(÷)	2.	(2 ÷) stored
3	())	0.6666667	(2 ÷ 3) evaluated
	())	8.3333333	(9 − (2 ÷ 3)) evaluated
	())	0.96	(8 ÷ (9 − (2 ÷ 3)))
	(=)	5.96	5 + (8 ÷ (9 − (2 ÷ 3)))

Because the (=) key completes all operations whenever it is used, it could have been used here instead of the three ()) keys. Try working this problem again and pressing (=) instead of the first ()).

Fig. 18.6. **Operations can be stored in the calculator's operating system.** *Courtesy Texas Instruments*

rithms, common logarithms and common antilogarithms, and trigonometric functions. The calculator can also be used to convert degrees to radians, degrees to grads, grads to degrees, grads to radians, radians to degrees, and radians to grads. The calculator can also be used to convert measurements in the English system of measurement to the metric system. The

Roots and Powers

(x^2) **Square Key**—Calculates the square of the number x in the display.

Example: $(4.235)^2 = 17.935225$

Enter	Press	Display
4.235	(x^2)	17.935225

($\sqrt{x}$) **Square Root Key**—Calculates the square root of the number x in the display. The x value cannot be negative.

Example: $\sqrt{6.25} = 2.5$

Enter	Press	Display
6.25	($\sqrt{x}$)	2.5

Fig. 18.7. **Roots and powers keys can easily be used to solve complex problems.** *Courtesy Texas Instruments*

Conversion Factors

English to Metric

To Find	Multiply	By
microns	mils	**25.4**
centimeters	inches	**2.54**
meters	feet	**0.3048**
meters	yards	**0.9144**
kilometers	miles	**1.609344**
gram	ounces	28.349523
kilogram	pounds	4.5359237×10^{-1}
liters	gallons (U.S.)	3.7854118
liters	gallons (Imp.)	4.546090
milliliters (cc)	fl. ounces	29.573530
sq. centimeters	sq. inches	**6.4516**
sq. meters	sq. feet	9.290304×10^{-2}
sq. meters	sq. yards	8.3612736×10^{-1}
milliliters (cc)	cu. inches	**16.387064**
cu. meters	cu. feet	2.8316847×10^{-2}
cu. meters	cu. yards	7.6455486×10^{-1}

Boldface numbers are exact; others are rounded.

Temperature Conversions

$$°F = \frac{9}{5}(°C) + 32$$
$$°C = \frac{5}{9}(°F - 32)$$

Fig. 18.8. The calculator can be used to convert drawing measurements from the English to the metric system with the use of this chart. *Courtesy Texas Instruments*

conversion factors are illustrated in figure 18.8. Temperature measurements can also be converted using the information in figure 18.8.

This type of calculator is valuable to those individuals who work with drawings in their everyday activities. The calculator allows an individual to handle many calculations accurately and quickly. This chapter has been designed to demonstrate the potential uses for the calculator in activities related to blueprints and drawings.

CHAPTER 19

Computer-Aided Design and Drafting

The use of computers is a cost effective method of increasing design and drafting productivity in mechanical, electronic, electrical, architectural, and civil engineering applications. Computer-aided design and drafting can increase productivity up to ten times with some of the systems that are available today (see fig. 19.1). Moreover, prices of these systems start at less than the yearly salary for one additional drafter. These computer-aided design and drafting systems help business and industry combat rapidly rising labor costs and manpower shortages.

What is a CAD/D system?

   Answer: A CAD/D system is a computer-aided design/ drafting system that generates drawings from computer programs.

What are the advantages of the CAD/D system?

   Answer: It is a cost effective means of increasing design and drafting productivity for mechanical, electronic/electrical, architectural, and civil engineering applications as well as other types of activities that require extensive use of drawings. It is also an effective solution to rapidly rising labor costs and manpower shortages.

What is an on-line interactive tutorial system?

   Answer: It acquaints the user of the CAD/D system with the various components and capability of the system. It also provides instruction on every major command and feature of the system.

How much time is required before you can begin drawing with the system?

Fig. 19.1. A computer-aided design and drafting system. *Courtesy T and W Systems, Inc.*

Answer: Once the training program has been completed the designer/drafter can begin drawing by relying on the prompts and commands built into the system (see fig. 19.2).

What are some of the major advantages of the language employed in the CAD/D system?

Answer: It can be easily learned and used by drafters/designers. The language also allows the user to tailor the system to specific application requirements without system reprogramming.

Why is it an advantage to visualize parts in three dimensions?

Answer: Engineers and designers in a non-CAD/D environment think in three dimensions but draw in two dimensions. The advantage of visualizing a part in three dimensions is that time, money, and rework time will be saved. CAD/D programs can provide truly automated three-dimensional drawing.

Is it possible to have various line widths and line colors with a CAD/D system?

Answer: CAD/D systems have the capability to produce different width lines and also lines in different colors.

What are some examples of comprehensive graphics editing capabilities of CAD/D systems?

Answer: These editing systems have the capabilities to

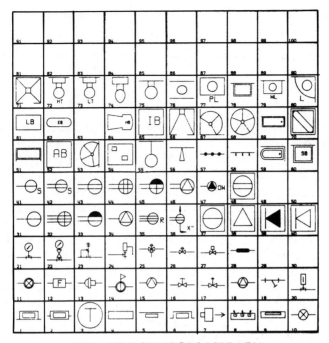

EXAMPLE SYMBOLS LIBRARY

Fig. 19.2. A symbols library for use with a computer-aided design and drafting system. *Courtesy T and W Systems, Inc.*

move or remove line segments, line segments that intersect, portions of line segments, and endpoints. These are examples of some of the capabilities of such systems.

What is the text-processing capability of CAD/D systems?

Answer: The text-processing capability of these systems allows for the convenient processing of text and annotations to be merged into drawings generated by the system.

What is the text-merge feature found in some CAD/D systems?

Answer: The text-merge feature, which is related to the text-processing capability, allows the text that has been created to be inserted into the drawings. This is a useful function where there is a large amount of tabular data that must be included in drawings. It is also useful where the text has been generated by other applications. The text-merge feature also allows the text and annotated material to be displayed in a number of

295

different standard type fonts or any number of user defined fonts.

Do these systems work by themselves or with a large number of other systems?

Answer: These systems can be operated as standalone operations or as part of a network of systems.

What are some of the automatic features that can be found in CAD/ D systems?

Answer: These systems have a number of different automatic features that are employed to make the generation of accurate drawings more efficient. There are a number of automatic utilities, including automatic cross hatching using user-defined and predefined patterns, double line options, fillets, splines (curve fitting), and semiautomatic dimensioning. With the semiautomatic dimensioning capability, the user can automatically generate linear dimensions, including extension lines, leaders dimension lines, and annotation by pointing at the drawing feature. Dimensions can also be entered manually or automatically in decimal notation, or in feet, inches, and fractions (see fig. 19.3).

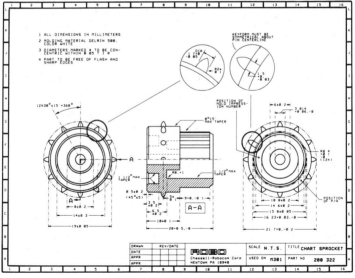

Fig. 19.3. A computer-generated drawing with all dimensions in the metric system. *Courtesy ROBO Systems*

What about the bill of material capacity in these systems?

Answer: Some of these systems have a sophisticated bill of material capability that will allow the user to associate text information with predefined figures and symbols used in drawings so that a complete bill of material can be produced with one simple command. The text information generated for the bill of materials includes quantities, cost, part numbers, and colors (see fig. 19.4).

What type of accounting is used to calculate the time involved on various drawing and design assignments?

Answer: Integral job accounting is used. This type of accounting allows the system to keep running totals of both wall clock and CPU time categorized by either drawing name, job number, work order, operator, or activity. The systems are so designed that accounting can be done for each work station or plotter.

What about the multitasking capability of these systems?

Answer: The multitasking capability allows plotting, editing, communications, and other tasks to be performed concurrently.

How are text and graphic files handled in these systems?

Answer: Text and graphic files can be easily loaded out to and in from storage without difficulty (see fig. 19.5).

What types of geometric construction aids are available in these systems?

Answer: There are a number of geometric construction aids that facilitate the creation of parallel and perpendicular lines, tangents to one or more circles, compute line length, extend lines, arc radius, angle between two line, areas, and perimeters.

What about the menu and menu-page capability?

Answer: Some of the more advanced systems have menu and menu-page capability that will help the user define special command menus to supplement the standard menu. This capability is extremely important in large drafting departments because a drafting department can define a separate menu for each different type of drawing need.

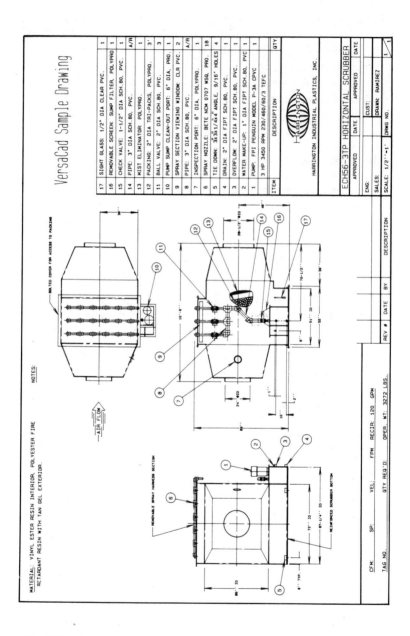

Fig. 19.4. A computer-generated drawing with a complete bill of materials keyed to parts numbers. *Courtesy T and W Systems, Inc.*

Fig. 19.5. A software package for a computer-aided design and drafting system. *Courtesy ROBO Systems*

What are advantages of CAD/D systems that are interfaced with other CAD/D systems?

  Answer: Data can be exchanged with other systems so that each system can perform its most efficient role in the hierarchy of CAD/D processing.

What is utility management in CAD/D systems?

  Answer: Once the line work and annotation phase of the drawing is complete, several utility commands can be used to perform a number of different functions that include saving or loading a drawing, plotting or transmitting a drawing, or clearing a drawing from the screen. Many drawing-oriented utilities included in these systems perform the following functions:

Where to locate point coordinates

Grid to display a reference grid of various scales

Units for converting English to Metric and vice versa

Length to measure distance

Area to calculate enclosed areas

Display to view the entire drawing or enlarged areas of the drawing

What types of modification commands are incorporated into these systems?

Answer: Some systems have design/drawing modification commands that are provided to rotate, translate, or scale figures in the drawing. If a symbol is used repetitively, for example, a resistor or a bolt, it can be scaled, rotated, or translated prior to its insertion into the drawing. In addition, objects can be scaled at different values in the x, y, and z coordinate axes. There are other modifications that can also be used frequently, such as selective pen use when plotting, transformation, and paper distortion correction. This last feature is helpful in digitizing an existing drawing.

What types of design-editing features are incorporated into these systems?

Answer: Most of these systems have a complete design-editing facility so that fast easy changes can be made to drawings. A number of different commands are utilized in design editing to speed the completion of the corrected drawing. For example, *backspace, delete, edit, stretch edit,* and *move edit* are a few of the commands that are employed in this activity. With these commands the user can rapidly delete, stretch or move any or all selected lines, text, or figures within the original drawing. The systems also protect the operator from unwanted editing results by automatically creating a copy of the drawing before the editing function is started. If the operator wants the original drawing to remain unchanged, the new drawing, which can be created as the result of editing, may be saved under a

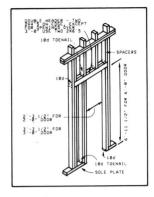

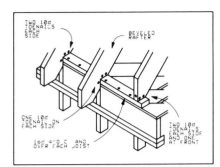

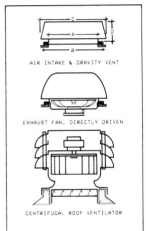

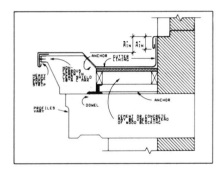

Fig. 19.6. Computer-generated detail drawings used in the construction industry. *Courtesy ROBO Systems*

different name. Thus, two versions of the same drawing will be readily available when needed.

What are some special features of CAD/D systems?

Answer: Frequently used commands designed to do various parts of drawings can be grouped into predetermined sequences. These predetermined sequences, when executed, will perform multiple functions by the entry of a single command.

What are some of the applications for CAD/D systems?

Answer: CAD/D systems can be used in a number of different activities that require multiple drawings on a continual basis. CAD/D systems are now used extensively in mechanical engineering, electrical engineering, electronic drafting, architectural drawings, piping drawings, civil engineering, facilities planning, maintenance drawings, project control, and in many other applications (see fig. 19.6).

APPENDIX

MISCELLANEOUS MEASUREMENTS

Measures of Length

1 mile = 1760 yards = 5280 feet.
1 yard = 3 feet = 36 inches.
1 foot = 12 inches.
The following measures of length are also used occasionally:
1 mil = 0.001 inch.   1 fathom = 2 yards = 6 feet.
1 rod = 5.5 yards = 16.5 feet.   1 hand = 4 inches.
1 span = 9 inches.

Surveyor's Measure

1 mile = 8 furlongs = 80 chains.
1 furlong = 10 chains = 220 yards.
1 chain = 4 rods = 22 yards = 66 feet = 100 links.
1 link = 7.92 inches.

Square Measure

1 square mile = 640 acres = 6400 square chains.
1 acre = 10 square chains = 4840 square yards = 43,560 square feet.
1 square chain = 16 square rods = 484 square yards = 4356 square feet.
1 square rod = 30.25 square yards = 272.25 square feet = 625 square links.
1 square yard = 9 square feet.
1 square foot = 144 square inches.
An acre is equal to a square, the side of which is 208.7 feet.

THE METRIC SYSTEM

Measures of Length

1 Millimeter (mm.) = ...0.03937079 inch, or about 1/25 inch
10 Millimeters = 1 Centimeter (cm.) = ...0.3937079 inch
10 Centimeters = 1 Decimeter (dm.) = ...3.937079 inch
10 Decimeters = 1 Meter (m.) = 39.37079 inches, 3.2808992 feet or 1.09361 yards
10 Meters = 1 Decameter (Dm.) = ...32.808992 feet
10 Decameters = 1 Hectometer (Hm.) = ...19.927817 rods
10 Hectometers = 1 Kilometer (Km.) = 1093.61 yards or 0.6213824 mile
10 Kilometers = 1 Myriameter (Mm.) = ...6.213824 miles
1 inch = 2.54 cm., 1 foot = 0.3048 m., 1 yard = 0.9144 m., 1 rod = 0.5029 Dm.,
     1 mile = 1.6093 Km.

Courtesy L.S. Starrett Co.

ENGLISH TO METRIC CONVERSION TABLE

Decimals to Millimeters

Decimal	mm	Decimal	mm
0.001	0.0254	0.500	12.7000
0.002	0.0508	0.510	12.9540
0.003	0.0762	0.520	13.2080
0.004	0.1016	0.530	13.4620
0.005	0.1270	0.540	13.7160
0.006	0.1524	0.550	13.9700
0.007	0.1778	0.560	14.2240
0.008	0.2032	0.570	14.4780
0.009	0.2286	0.580	14.7320
		0.590	14.9860
0.010	0.2540		
0.020	0.5080		
0.030	0.7620		
0.040	1.0160	0.600	15.2400
0.050	1.2700	0.610	15.4940
0.060	1.5240	0.620	15.7480
0.070	1.7780	0.630	16.0020
0.080	2.0320	0.640	16.2560
0.090	2.2860	0.650	16.5100
		0.660	16.7640
0.100	2.5400	0.670	17.0180
0.110	2.7940	0.680	17.2720
0.120	3.0480	0.690	17.5260
0.130	3.3020		
0.140	3.5560		
0.150	3.8100		
0.160	4.0640	0.700	17.7800
0.170	4.3180	0.710	18.0340
0.180	4.5720	0.720	18.2880
0.190	4.8260	0.730	18.5420
		0.740	18.7960
0.200	5.0800	0.750	19.0500
0.210	5.3340	0.760	19.3040
0.220	5.5880	0.770	19.5580
0.230	5.8420	0.780	19.8120
0.240	6.0960	0.790	20.0660
0.250	6.3500		
0.260	6.6040		
0.270	6.8580		
0.280	7.1120	0.800	20.3200
0.290	7.3660	0.810	20.5740
		0.820	20.8280
0.300	7.6200	0.830	21.0820
0.310	7.8740	0.840	21.3360
0.320	8.1280	0.850	21.5900
0.330	8.3820	0.860	21.8440
0.340	8.6360	0.870	22.0980
0.350	8.8900	0.880	22.3520
0.360	9.1440	0.890	22.6060
0.370	9.3980		
0.380	9.6520		
0.390	9.9060		
		0.900	22.8600
0.400	10.1600	0.910	23.1140
0.410	10.4140	0.920	23.3680
0.420	10.6680	0.930	23.6220
0.430	10.9220	0.940	23.8760
0.440	11.1760	0.950	24.1300
0.450	11.4300	0.960	24.3840
0.460	11.6840	0.970	24.6380
0.470	11.9380	0.980	24.8920
0.480	12.1920	0.990	25.1460
0.490	12.4460	1.000	25.4000

Fractions to Decimals to Millimeters

Fraction	Decimal	mm	Fraction	Decimal	mm
1/64	0.0156	0.3969	33/64	0.5156	13.0969
1/32	0.0312	0.7938	17/32	0.5312	13.4938
3/64	0.0469	1.1906	35/64	0.5469	13.8906
1/16	0.0625	1.5875	9/16	0.5625	14.2875
5/64	0.0781	1.9844	37/64	0.5781	14.6844
3/32	0.0938	2.3812	19/32	0.5938	15.0812
7/64	0.1094	2.7781	39/64	0.6094	15.4781
1/8	0.1250	3.1750	5/8	0.6250	15.8750
9/64	0.1406	3.5719	41/64	0.6406	16.2719
5/32	0.1562	3.9688	21/32	0.6562	16.6688
11/64	0.1719	4.3656	43/64	0.6719	17.0656
3/16	0.1875	4.7625	11/16	0.6875	17.4625
13/64	0.2031	5.1594	45/64	0.7031	17.8594
7/32	0.2188	5.5562	23/32	0.7188	18.2562
15/64	0.2344	5.9531	47/64	0.7344	18.6531
1/4	0.2500	6.3500	3/4	0.7500	19.0500
17/64	0.2656	6.7469	49/64	0.7656	19.4469
9/32	0.2812	7.1438	25/32	0.7812	19.8438
19/64	0.2969	7.5406	51/64	0.7969	20.2406
5/16	0.3125	7.9375	13/16	0.8125	20.6375
21/64	0.3281	8.3344	53/64	0.8281	21.0344
11/32	0.3438	8.7312	27/32	0.8438	21.4312
23/64	0.3594	9.1281	55/64	0.8594	21.8281
3/8	0.3750	9.5250	7/8	0.8750	22.2250
25/64	0.3906	9.9219	57/64	0.8906	22.6219
13/32	0.4062	10.3188	29/32	0.9062	23.0188
27/64	0.4219	10.7156	59/64	0.9219	23.4156
7/16	0.4375	11.1125	15/16	0.9375	23.8125
29/64	0.4531	11.5094	61/64	0.9531	24.2094
15/32	0.4688	11.9062	31/32	0.9688	24.6062
31/64	0.4844	12.3031	63/64	0.9844	25.0031
1/2	0.5000	12.7000	1	1.0000	25.4000

Courtesy L.S. Starrett Co.

ALLOWANCES FOR DIFFERENT CLASSES OF FIT

Class	Nominal Diameters	Tolerances in Standard Holes*					
		Up to ½"	9/16"-1"	1 1/16"-2"	2 1/16"-3"	3 1/16"-4"	4 1/16"-5"
A	High Limit	+ 0.00025	+ 0.0005	+ 0.00075	+ 0.0010	+ 0.0010	+ 0.0010
	Low Limit	− 0.00025	− 0.00025	− 0.00025	− 0.0005	− 0.0005	− 0.0005
	Tolerance	0.0005	0.00075	0.0010	0.0015	0.0015	0.0015
B	High Limit	+ 0.0005	+ 0.00075	+ 0.0010	+ 0.00125	+ 0.0015	+ 0.00175
	Low Limit	− 0.0005	− 0.0005	− 0.0005	− 0.00075	− 0.00075	− 0.00075
	Tolerance	0.0010	0.00125	0.0015	0.0020	0.00225	0.0025
Allowances for Forced Fits							
F	High Limit	+ 0.0010	+ 0.0020	+ 0.0040	+ 0.0060	+ 0.0080	+ 0.0100
	Low Limit	+ 0.0005	+ 0.0015	+ 0.0030	+ 0.0045	+ 0.0060	+ 0.0080
	Tolerance	0.0005	0.0005	0.0010	0.0015	0.0020	0.0020
Allowances for Driving Fits							
D	High Limit	+ 0.0005	+ 0.0010	+ 0.0015	+ 0.0025	+ 0.0030	+ 0.0035
	Low Limit	+ 0.00025	+ 0.00075	+ 0.0010	+ 0.0015	+ 0.0020	+ 0.0025
	Tolerance	0.00025	0.00025	0.0005	0.0010	0.0010	0.0010
Allowances for Push Fits							
P	High Limit	− 0.00025	− 0.00025	− 0.00025	− 0.0005	− 0.0005	− 0.0005
	Low Limit	− 0.00075	− 0.00075	− 0.00075	− 0.0010	− 0.0010	− 0.0010
	Tolerance	0.0005	0.0005	0.0005	0.0005	0.0005	0.0005
Allowances for Running Fits†							
X	High Limit	− 0.0010	− 0.00125	− 0.00175	− 0.0020	− 0.0025	− 0.0030
	Low Limit	− 0.0020	− 0.00275	− 0.0035	− 0.00425	− 0.0050	− 0.00575
	Tolerance	0.0010	0.0015	0.00175	0.00225	0.0025	0.00275
Y	High Limit	− 0.00075	− 0.0010	− 0.00125	− 0.0015	− 0.0020	− 0.00225
	Low Limit	− 0.00125	− 0.0020	− 0.0025	− 0.0030	− 0.0035	− 0.0040
	Tolerance	0.0005	0.0010	0.00125	0.0015	0.0015	0.00175
Z	High Limit	− 0.0005	− 0.00075	− 0.00075	− 0.0010	− 0.0010	− 0.00125
	Low Limit	− 0.00075	− 0.00125	− 0.0015	− 0.0020	− 0.00225	− 0.0025
	Tolerance	0.00025	0.0005	0.00075	0.0010	0.00125	0.00125

* Tolerance is provided for holes, which ordinary standard reamers can produce, in two grades, Classes A and B. the selection of which is a question for the user's decision and dependent upon the quality of the work required; some prefer to use Class A as working limits and Class B as inspection limits.

† Running fits, which are the most commonly required, are divided into three grades: Class X for engine and other work where easy fits are wanted; Class Y for high speeds and good average machine work; Class Z for fine tool work.

Courtesy L.S. Starrett Co.

DECIMAL EQUIVALENTS, SQUARES, CUBES, SQUARE AND CUBE ROOTS, CIRCUMFERENCES AND AREAS OF CIRCLES, FROM 1/64 to 5/8 INCH

Frac-tion	Dec. Equiv.	Square	Sq. Root	Cube	Cube Root	Circle*	
						Circum.	Area
1/64	.015625	.0002441	.125	.000003815	.25	.04909	.000192
1/32	.03125	.0009766	.176777	.000030518	.31498	.09817	.000767
3/64	.046875	.0021973	.216506	.000102997	.36056	.14726	.001726
1/16	.0625	.0039063	.25	.00024414	.39685	.19635	.003068
5/64	.078125	.0061035	.279508	.00047684	.42749	.24544	.004794
3/32	.09375	.0087891	.306186	.00082397	.45428	.29452	.006903
7/64	.109375	.0119629	.330719	.0013084	.47823	.34361	.009396
1/8	.125	.015625	.353553	.0019531	.5	.39270	.012272
9/64	.140625	.0197754	.375	.0027809	.52002	.44179	.015532
5/32	.15625	.0244141	.395285	.0038147	.53861	.49087	.019175
11/64	.171875	.0295410	.414578	.0050774	.55600	.53996	.023201
3/16	.1875	.0351563	.433013	.0065918	.57236	.58905	.027611
13/64	.203125	.0412598	.450694	.0083809	.58783	.63814	.032405
7/32	.21875	.0478516	.467707	.010468	.60254	.68722	.037583
15/64	.234375	.0549316	.484123	.012875	.61655	.73631	.043143
1/4	.25	.0625	.5	.015625	.62996	.78540	.049087
17/64	.265625	.0705566	.515388	.018742	.64282	.83449	.055415
9/32	.28125	.0791016	.530330	.022247	.65519	.88357	.062126
19/64	.296875	.0881348	.544862	.026165	.66710	.93266	.069221
5/16	.3125	.0976562	.559017	.030518	.67860	.98175	.076699
21/64	.328125	.107666	.572822	.035328	.68973	1.03084	.084561
11/32	.34375	.118164	.586302	.040619	.70051	1.07992	.092806
23/64	.359375	.129150	.599479	.046413	.71097	1.12901	.101434
3/8	.375	.140625	.612372	.052734	.72112	1.17810	.110445
25/64	.390625	.1525879	.625	.059605	.73100	1.22718	.119842
13/32	.40625	.1650391	.637377	.067047	.74062	1.27627	.129621
27/64	.421875	.1779785	.649519	.075085	.75	1.32536	.139784
7/16	.4375	.1914063	.661438	.083740	.75915	1.37445	.150330
29/64	.453125	.2053223	.673146	.093037	.76808	1.42353	.161260
15/32	.46875	.2197266	.684653	.102997	.77681	1.47262	.172573
31/64	.484375	.2346191	.695971	.113644	.78535	1.52171	.184269
1/2	.5	.25	.707107	.125	.79370	1.57080	.196350
33/64	.515625	.265869	.718070	.137089	.80188	1.61988	.208813
17/32	.53125	.282227	.728869	.149933	.80990	1.66897	.221660
35/64	.546875	.299072	.739510	.163555	.81777	1.71806	.234891
9/16	.5625	.316406	.75	.177979	.82548	1.76715	.248505
37/64	.578125	.334229	.760345	.193226	.83306	1.81623	.262502
19/32	.59375	.352539	.770552	.209320	.84049	1.86532	.276884
39/64	.609375	.371338	.780625	.226284	.84780	1.91441	.291648
5/8	.625	.390625	.790569	.244141	.85499	1.96350	.306796

* Fraction represents diameter.

continued on next page

DECIMAL EQUIVALENTS, SQUARES, CUBES, SQUARE AND CUBE ROOTS, CIRCUMFERENCES AND AREAS OF CIRCLES, FROM $^{41}/_{64}$ to 1 INCH

Frac-tion	Dec. Equiv.	Square	Sq. Root	Cube	Cube Root	Circle* Circum.	Circle* Area
$^{41}/_{64}$	.640625	.410400	.800391	.262913	.86205	2.01258	.322328
$^{21}/_{32}$	.65625	.430664	.810093	.282623	.86901	2.06167	.338243
$^{43}/_{64}$	.671875	.451416	.819680	.303295	.87585	2.11076	.354541
$^{11}/_{16}$	.6875	.472656	.829156	.324951	.88259	2.15984	.371223
$^{45}/_{64}$	.703125	.494385	.838525	.347614	.88922	2.20893	.388289
$^{23}/_{32}$	.71875	.516602	.847791	.371307	.89576	2.25802	.405737
$^{47}/_{64}$	.734375	.539307	.856957	.396053	.90221	2.30711	.423570
$^{3}/_{4}$	.75	.5625	.866025	.421875	.90856	2.35619	.441786
$^{49}/_{64}$	.765625	.586182	.875	.448795	.91483	2.40528	.460386
$^{25}/_{32}$	.78125	.610352	.883883	.476837	.92101	2.45437	.479369
$^{51}/_{64}$	.796875	.635010	.892679	.506023	.92711	2.50346	.498736
$^{13}/_{16}$	.8125	.660156	.901388	.536377	.93313	2.55254	.518486
$^{53}/_{64}$	.828125	.685791	.910014	.567921	.93907	2.60163	.538619
$^{27}/_{32}$	.84375	.711914	.918559	.600677	.94494	2.65072	.559136
$^{55}/_{64}$	.859375	.738525	.927024	.634670	.95074	2.69981	.580036
$^{7}/_{8}$	.875	.765625	.935414	.669922	.95647	2.74889	.601320
$^{57}/_{64}$	.890625	.793213	.943729	.706455	.96213	2.79798	.622988
$^{29}/_{32}$	.90625	.821289	.951972	.744293	.96772	2.84707	.645039
$^{59}/_{64}$	.921875	.849854	.960143	.783459	.97325	2.89616	.667473
$^{15}/_{16}$	.9375	.878906	.968246	.823975	.97872	2.94524	.690291
$^{61}/_{64}$	.953125	.908447	.976281	.865864	.98412	2.99433	.713493
$^{31}/_{32}$	.96875	.938477	.984251	.909149	.98947	3.04342	.737078
$^{63}/_{64}$	.984375	.968994	.992157	.953854	.99476	3.09251	.761046
1	1	1	1	1	1	3.14159	.785398

* Fraction represents diameter.

Courtesy L.S. Starrett Co.

SCREW THREADS AND TAP DRILL SIZES

N C or A. S. M. E. SPECIAL MACHINE SCREWS

Size of Tap	Thds. per inch	Tap Drill	Body Drill
1	64	53	48
2	56	50	44
3	48	47	39
4	40	43	33
5	40	38	$^{1}/_{8}$
6	32	36	28
8	32	29	19
10	24	25	11
12	24	16	$^{7}/_{32}$

N F or A. S. M. E. SPECIAL MACHINE SCREWS

Size of Tap	Thds. per inch	Tap Drill	Body Drill
2	64	50	44
3	56	45	39
4	48	42	33
5	44	37	$^{1}/_{8}$
6	40	33	28
8	36	29	19
10	32	21	11
*10	30	22	11
12	28	14	$^{7}/_{32}$

* A.S.M.E. Only

AMERICAN STANDARD TAPER PIPE THREADS

Size of Tap	Thds. per inch	Tap Drill
$^{1}/_{8}$	27	$^{11}/_{32}$
$^{1}/_{4}$	18	$^{7}/_{16}$
$^{3}/_{8}$	18	$^{19}/_{32}$
$^{1}/_{2}$	14	$^{23}/_{32}$
$^{3}/_{4}$	14	$^{15}/_{16}$
1	$11^{1}/_{2}$	$1 \, ^{5}/_{32}$
$1^{1}/_{4}$	$11^{1}/_{2}$	$1 \, ^{1}/_{2}$
$1^{1}/_{2}$	$11^{1}/_{2}$	$1 \, ^{23}/_{32}$
2	$11^{1}/_{2}$	$2 \, ^{3}/_{16}$
$2^{1}/_{2}$	8	$2 \, ^{5}/_{8}$
3	8	$3 \, ^{1}/_{4}$

N F or S.A.E. STANDARD SCREWS

Size of Tap	Thds. per inch	Tap Drill
$^{1}/_{4}$	28	3
$^{5}/_{16}$	24	1
$^{3}/_{8}$	24	0
$^{7}/_{16}$	20	$^{25}/_{64}$
$^{1}/_{2}$	20	$^{29}/_{64}$
$^{9}/_{16}$	18	$^{33}/_{64}$
$^{5}/_{8}$	18	$^{37}/_{64}$
$^{11}/_{16}$	16	$^{5}/_{8}$
$^{3}/_{4}$	16	$^{11}/_{16}$
$^{7}/_{8}$	14	$^{13}/_{16}$
.1	14	$^{15}/_{16}$
$1 \, ^{1}/_{8}$	12	$1 \, ^{3}/_{64}$

Tap Drills allow approx. 75% Full Thread

Courtesy L.S. Starrett Co.

NUMBER AND LETTER SIZES OF DRILLS
WITH DECIMAL EQUIVALENTS

Sizes starting with No. 80 and going up to 1 inch. This table is useful for quickly determining the nearest drill size for any decimal, for root diameters, body drills, etc.

Drill No.	Frac.	Deci.	Drill No.	Frac.	Deci.	Drill No.	Frac.	Deci.	Drill No.	Frac.	Deci.
80	—	.0135	42	—	.0935	7	—	.201	X	—	.397
79	—	.0145	—	3/32	.0938	—	13/64	.203	Y	—	.404
—	1/64	.0156				6	—	.204	—	13/32	.406
78	—	.0160	41	—	.0960	5	—	.206	Z	—	.413
77	—	.0180	40	—	.0980	4	—	.209	—	27/64	.422
			39	—	.0995						
76	—	.0200	38	—	.1015	3	—	.213	—	7/16	.438
75	—	.0210	37	—	.1040	—	7/32	.219	—	29/64	.453
74	—	.0225				2	—	.221			
73	—	.0240	36	—	.1065	1	—	.228	—	15/32	.469
72	—	.0250	—	7/64	.1094	A	—	.234	—	31/64	.484
			35	—	.1100				—	1/2	.500
71	—	.0260	34	—	.1110	—	15/64	.234	—	33/64	.516
70	—	.0280	33	—	.1130	B	—	.238	—	17/32	.531
69	—	.0292				C	—	.242			
68	—	.0310	32	—	.116	D	—	.246	—	35/64	.547
—	1/32	.0313	31	—	.120	—	1/4	.250	—	9/16	.562
			—	1/8	.125				—	37/64	.578
67	—	.0320	30	—	.129	E	—	.250	—	19/32	.594
66	—	.0330	29	—	.136	F	—	.257	—	39/64	.609
65	—	.0350				G	—	.261			
64	—	.0360	—	9/64	.140	—	17/64	.266	—	5/8	.625
63	—	.0370	28	—	.141	H	—	.266	—	41/64	.641
			27	—	.144				—	21/32	.656
62	—	.0380	26	—	.147	I	—	.272	—	43/64	.672
61	—	.0390	25	—	.150	J	—	.277	—	11/16	.688
60	—	.0400				—	9/32	.281			
59	—	.0410	24	—	.152	K	—	.281	—	45/64	.703
58	—	.0420	23	—	.154	L	—	.290	—	23/32	.719
			—	5/32	.156				—	47/64	.734
57	—	.0430	22	—	.157	M	—	.295	—	3/4	.750
56	—	.0465	21	—	.159	—	19/64	.297	—	49/64	.766
—	3/64	.0469				N	—	.302			
55	—	.0520				—	5/16	.313	—	25/32	.781
54	—	.0550	20	—	.161	O	—	.316	—	51/64	.797
			19	—	.166				—	13/16	.813
53	—	.0595	18	—	.170	P	—	.323	—	53/64	.828
—	1/16	.0625	—	11/64	.172	—	21/64	.328	—	27/32	.844
52	—	.0635	17	—	.173	Q	—	.332			
51	—	.0670				R	—	.339			
50	—	.0700	16	—	.177	—	11/32	.344	—	55/64	.859
			15	—	.180				—	7/8	.875
49	—	.0730	14	—	.182	S	—	.348	—	57/64	.891
48	—	.0760	13	—	.185	T	—	.358	—	29/32	.906
—	5/64	.0781	—	3/16	.188	—	23/64	.359	—	59/64	.922
47	—	.0785				U	—	.368			
46	—	.0810	12	—	.189	—	3/8	.375	—	15/16	.938
			11	—	.191				—	61/64	.953
45	—	.0820	10	—	.194	V	—	.377	—	31/32	.969
44	—	.0860	9	—	.196	W	—	.386	—	63/64	.984
43	—	.0890	8	—	.199	—	25/64	.391	—	1	1.000

Courtesy L.S. Starrett Co.

DOUBLE DEPTH OF THREADS

Threads per In. N	V Threads D D	Am. Nat. Form DD U. S. Std.	Whitworth Standard D D	Threads per In. N	V Threads D D	Am. Nat. Form DD U. S. Std.	Whitworth Standard D D
2	.86650	.64950	.64000	28	.06185	.04639	.04571
2¹/₄	.77022	.57733	.56888	30	.05773	.04330	.04266
2³/₈	.72960	.45694	.53894	32	.05412	.04059	.04000
2¹/₂	.69320	.51960	.51200	34	.05097	.03820	.03764
2⁵/₈	.66015	.49485	.48761	36	.04811	.03608	.03555
2³/₄	.63019	.47236	.45545	38	.04560	.03418	.03368
2⁷/₈	.60278	.45182	.44521	40	.04330	.03247	.03200
3	.57733	.43300	.42666	42	.04126	.03093	.03047
3¹/₄	.53323	.39966	.39384	44	.03936	.02952	.02909
3¹/₂	.49485	.37114	.35571	46	.03767	.02823	.02782
4	.43300	.32475	.32000	48	.03608	.02706	.02666
4¹/₂	.38438	.28869	.23444	50	.03464	.02598	.02560
5	.34660	.25980	.25600	52	.03332	.02498	.02461
5¹/₂	.31490	.23618	.23272	54	.03209	.02405	.02370
6	.28866	.21650	.21333	56	.03093	.02319	.02285
7	.24742	.18557	.13285	58	.02987	.02239	.02206
8	.21650	.16237	.15000	60	.02887	.02165	.02133
9	.19244	.14433	.14222	62	.02795	.02095	.02064
10	.17320	.12990	.12800	64	.02706	.02029	.02000
11	.15745	.11809	.11636	66	.02625	.01968	.01939
11¹/₂	.15069	.11295	.11121	68	.02548	.01910	.01882
12	.14433	.10825	.10666	70	.02475	.01855	.01728
13	.13323	.09992	.09846	72	.02407	.01804	.01782
14	.12357	.09278	.09142	74	.02341	.01752	.01729
15	.11555	.08660	.08533	76	.02280	.01714	.01673
16	.10825	.08118	.08000	78	.02221	.01665	.01641
18	.09622	.07216	.07111	80	.02166	.01623	.01600
20	.08660	.06495	.06400	82	.02113	.01584	.01560
22	.07872	.05904	.05818	84	.02063	.01546	.01523
24	.07216	.05412	.05333	86	.02015	.01510	.01476
26	.06661	.04996	.04923	88	.01957	.01476	.01454
27	.06418	.04811	.04740	90	.01925	.01443	.01422

$$D D = \frac{1.733}{N} \text{ For V Thread}$$

$$D D = \frac{1.299}{N} \text{ For American Nat. Form. U. S. Std.}$$

$$D D = \frac{1.28}{N} \text{ For Whitworth Standard}$$

Courtesy L.S. Starrett Co.

TAPERS AND ANGLES

Taper per Foot	Included Angle			Angle With Center Line			Taper per Inch	Taper per Inch from Center Line
	Deg.	Min.	Sec.	Deg.	Min.	Sec.		
$^1/_8$	0	35	47	0	17	54	.010416	.005208
$^3/_{16}$	0	53	44	0	26	52	.015625	.007812
$^1/_4$	1	11	38	0	35	49	.020833	.010416
$^5/_{16}$	1	29	31	0	44	46	.026042	.013021
$^3/_8$	1	47	25	0	53	42	.031250	.015625
$^7/_{16}$	2	5	18	1	2	39	.036458	.018229
$^1/_2$	2	23	12	1	11	36	.041667	.020833
$^9/_{16}$	2	41	7	1	20	34	.046875	.023438
$^5/_8$	2	59	3	1	29	31	.052084	.026042
$^{11}/_{16}$	3	16	56	1	38	28	.057292	.028646
$^3/_4$	3	34	48	1	47	24	.062500	.031250
$^{13}/_{16}$	3	52	42	1	56	21	.067708	.033854
$^7/_8$	4	10	32	2	5	16	.072917	.036456
$^{15}/_{16}$	4	28	26	2	14	13	.078125	.039063
1	4	46	19	2	23	10	.083330	.041667
1 $^1/_4$	5	57	45	2	58	53	.104166	.052084
1 $^1/_2$	7	9	10	3	34	35	.125000	.062500
1 $^3/_4$	8	20	28	4	10	14	.145833	.072917
2	9	31	37	4	45	49	.166666	.083332
2 $^1/_2$	11	53	38	5	56	49	.208333	.104166
3	14	2	0	7	1	0	.250000	.125000
3 $^1/_2$	16	35	39	8	17	49	.291666	.145833
4	18	55	31	9	27	44	.333333	.166666
4 $^1/_2$	21	14	20	10	37	10	.375000	.187500
5	23	32	12	11	46	6	.416666	.208333
6	28	4	20	14	2	10	.500000	.250000

Courtesy L.S. Starrett Co.

Index

The Audel®
Mail Order
Bookstore

Here's an opportunity to order the valuable books you may have missed before and to build your own personal, comprehensive library of Audel books. You can choose from an extensive selection of technical guides and reference books. They will provide access to the same sources the experts use, put all the answers at your fingertips, and give you the know-how to complete even the most complicated building or repairing job, in the same professional way.

Each volume:

- **Fully illustrated**
- **Packed with up-to-date facts and figures**
- **Completely indexed for easy reference**

APPLIANCES
HOME APPLIANCE SERVICING, 4th Edition
A practical book for electric & gas servicemen, mechanics & dealers. Covers the principles, servicing, and repairing of home appliances. 592 pages; $5\frac{1}{2} \times 8\frac{1}{4}$; hardbound. **Price: $15.95**

REFRIGERATION: HOME AND COMMERCIAL
Covers the whole realm of refrigeration equipment from fractional-horsepower water coolers through domestic refrigerators to multiton commercial installations. 656 pages; $5\frac{1}{2} \times 8\frac{1}{4}$; hardbound. **Price: $16.95**

AIR CONDITIONING: HOME AND COMMERCIAL
A concise collection of basic information, tables, and charts for those interested in understanding troubleshooting, and repairing home air-conditioners and commercial installations. 464 pages; $5\frac{1}{2} \times 8\frac{1}{4}$; hardbound. **Price: $14.95**

OIL BURNERS, 4th Edition
Provides complete information on all types of oil burners and associated equipment. Discusses burners—blowers—ignition transformers—electrodes—nozzles—fuel pumps—filters—controls. Installation and maintenance are stressed. 320 pages; $5\frac{1}{2} \times 8\frac{1}{4}$; hardbound. **Price: $12.95**

AUTOMOTIVE
AUTOMOBILE REPAIR GUIDE, 4th Edition
A practical reference for auto mechanics, servicemen, trainees, and owners. Explains theory, construction, and servicing of modern domestic motorcars. 800 pages; $5\frac{1}{2} \times 8\frac{1}{4}$; hardbound. **Price: $14.95**

Use the order coupon on the back of this book.
All prices are subject to change without notice.

AUTOMOTIVE AIR CONDITIONING

You can easily perform most all service procedures you've been paying for in the past. This book covers the systems built by the major manufacturers, even after-market installations. Contents: introduction—refrigerant—tools—air conditioning circuit—general service procedures—electrical systems—the cooling systems—system diagnosis—electrical diagnosis—troubleshooting. 232 pages; $5\frac{1}{2} \times 8\frac{1}{4}$; softcover. **Price: $7.95**

DIESEL ENGINE MANUAL, 4th Edition

A practical guide covering the theory, operation and maintenance of modern diesel engines. Explains diesel principles—valves—timing—fuel pumps—pistons and rings—cylinders—lubrication—cooling system—fuel oil and more. 480 pages; $5\frac{1}{2} \times 8\frac{1}{4}$; hardbound. **Price: $12.95**

GAS ENGINE MANUAL, 3rd Edition

A completely practical book covering the construction, operation, and repair of all types of modern gas engines. 400 pages; $5\frac{1}{2} \times 8\frac{1}{4}$; hardbound. **Price: $12.95**

SMALL GASOLINE ENGINES

A new manual providing practical and theoretical information for those who want to maintain and overhaul two- and four-cycle engines such as lawn mowers, edgers, snowblowers, outboard motors, electrical generators, and other equipment using engines up to 10 horsepower. 624 pp; $5\frac{1}{2} \times 8\frac{1}{4}$; hardbound. **Price: $15.95**

TRUCK GUIDE—3 Vols.

Three all-new volumes provide a primary source of practical information on truck operation and maintenance. Covers everything from basic principles (truck classification, construction components, and capabilities) to troubleshooting and repair. 1584 pages; $5\frac{1}{2} \times 8\frac{1}{4}$; hardbound. **Price: $41.85**
> **Volume 1**
> ENGINES: **$14.95**
> **Volume 2**
> ENGINE AUXILIARY SYSTEMS: **$14.95**
> **Volume 3**
> TRANSMISSIONS, STEERING AND BRAKES: **$14.95**

BUILDING AND MAINTENANCE
ANSWERS ON BLUEPRINT READING, 3rd Edition

Covers all types of blueprint reading for mechanics and builders. This book reveals the secret language of blueprints, step by step in easy stages. 312 pages; $5\frac{1}{2} \times 8\frac{1}{4}$; hardbound. **Price: $9.95**

BUILDING MAINTENANCE, 2nd Edition

Covers all the practical aspects of building maintenance. Painting and decorating; plumbing and pipe fitting; carpentry; heating maintenance; custodial practices and more. (A book for building owners, managers, and maintenance personnel.) 384 pages; $5\frac{1}{2} \times 8\frac{1}{4}$; hardbound. **Price: $9.95**

COMPLETE BUILDING CONSTRUCTION

At last—a one volume instruction manual to show you how to construct a frame or brick building from the footings to the ridge. Build your own garage, tool shed, other outbuildings—even your own house or place of business. Building construction tells you how to lay out the building and excavation lines on the lot; how to make concrete forms and pour the footings and foundation; how to make concrete slabs, walks, and driveways; how to lay concrete block, brick and tile; how to build your own fireplace and chimney. It's one of the newest Audel books, clearly written by experts in each field and ready to help you every step of the way. 800 pages; $5\frac{1}{2} \times 8\frac{1}{4}$; hardbound. **Price: $19.95**

Use the order coupon on the back of this book.
All prices are subject to change without notice.

GARDENING, LANDSCAPING, & GROUNDS MAINTENANCE, 3rd Edition

A comprehensive guide for homeowners and for industrial, municipal, and estate grounds-keepers. Gives information on proper care of annual and perennial flowers; various house plants; greenhouse design and construction; insect and rodent controls; and more. 416 pages; 5½ × 8¼; hardbound. **Price: $15.95**

CARPENTERS & BUILDERS LIBRARY, 5th Edition (4 Vols.)

A practical, illustrated trade assistant on modern construction for carpenters, builders, and all woodworkers. Explains in practical, concise language and illustrations all the principles, advances, and shortcuts based on modern practice. How to calculate various jobs. **Price: $39.95**

Volume 1
Tools, steel square, saw filing, joinery cabinets. 384 pages; 5½ × 8¼; hardbound. **Price: $10.95**

Volume 2
Mathematics, plans, specifications, estimates. 304 pages; 5½ × 8¼; hardbound. **Price: $10.95**

Volume 3
House and roof framing, layout foundations. 304 pages; 5½ × 8¼; hardbound. **Price: $10.95**

Volume 4
Doors, windows, stairs, millwork, painting. 368 pages; 5½ × 8¼; hardbound. **Price: $10.95**

HEATING, VENTILATING, AND AIR CONDITIONING LIBRARY (3 Vols.)

This three-volume set covers all types of furnaces, ductwork, air conditioners, heat pumps, radiant heaters, and water heaters, including swimming-pool heating systems. **Price: $41.95**

Volume 1
Partial Contents: Heating Fundamentals—Insulation Principles—Heating Fuels—Electric Heating System—Furnace Fundamentals—Gas-Fired Furnaces—Oil-Fired Furnaces—Coal-Fired Furnaces—Electric Furnaces. 614 pages; 5½ × 8¼; hardbound. **Price: $14.95**

Volume 2
Partial Contents: Oil Burners—Gas Burners—Thermostats and Humidistats—Gas and Oil Controls—Pipes, Pipe Fitting, and Piping Details—Valves and Valve Installations. 560 pages; 5½ × 8¼; hardbound. **Price: $14.95**

Volume 3
Partial Contents: Radiant Heating—Radiators, Convectors, and Unit Heaters—Stoves, Fireplaces, and Chimneys—Water Heaters and Other Appliances—Central Air Conditioning Systems—Humidifiers and Dehumidifiers. 544 pages; 5½ × 8¼; hardbound. **Price: $14.95**

HOME-MAINTENANCE AND REPAIR: Walls, Ceilings, and Floors

Easy-to-follow instructions for sprucing up and repairing the walls, ceiling, and floors of your home. Covers nail pops, plaster repair, painting, paneling, ceiling and bathroom tile, and sound control. 80 pages; 8½ × 11; softcover. **Price: $6.95**

HOME PLUMBING HANDBOOK, 3rd Edition

A complete guide to home plumbing repair and installation, 200 pages; 8½ × 11; softcover. **Price: $8.95**

MASONS AND BUILDERS LIBRARY, 2nd Edition—2 Vols.

A practical, illustrated trade assistant on modern construction for bricklayers, stonemasons, cement workers, plasterers, and tile setters. Explains all the principles, advances, and shortcuts based on modern practice—including how to figure and calculate various jobs. **Price: $24.90**

Volume 1
Concrete Block, Tile, Terrazzo. 368 pages; 5½ × 8¼; hardbound. **Price: $12.95**

Use the order coupon on the back of this book.
All prices are subject to change without notice.

Volume 2
Bricklaying, Plastering Rock Masonry, Clay Tile. 384 pages; 5½ × 8¼; hardbound.
Price: $12.95

PAINTING AND DECORATING

This all-inclusive guide to the principles and practice of coating and finishing interior and exterior surfaces is a fundamental sourcebook for the working painter and decorator and an invaluable guide for the serious amateur or building owner. Provides detailed descriptions of materials, pigmenting and mixing procedures, equipment, surface preparation, restoration, repair, and antiquing of all kinds of surfaces. 608 pages; 5½ × 8¼; hardbound. **Price: $18.95**

PLUMBERS AND PIPE FITTERS LIBRARY, 3rd Edition—3 Vols.

A practical, illustrated trade assistant and reference for master plumbers, journeymen and apprentice pipe fitters, gas fitters and helpers, builders, contractors, and engineers. Explains in simple language, illustrations, diagrams, charts, graphs, and pictures the principles of modern plumbing and pipe-fitting practices. **Price: $32.85**

Volume 1
Materials, tools, roughing-in. 320 pages; 5½ × 8¼; hardbound. **Price: $11.95**
Volume 2
Welding, heating, air-conditioning. 384 pages; 5½ × 8¼; hardbound. **Price: $11.95**
Volume 3
Water supply, drainage, calculations. 272 pages; 5½ × 8¼; hardbound. **Price: $11.95**

THE PLUMBERS HANDBOOK, 7th Edition

A pocket manual providing reference material for plumbers and/or pipe fitters. General information sections contain data on cast-iron fittings, copper drainage fittings, plastic pipe, and repair of fixtures. 330 pages; 4 × 6 softcover. **Price: $9.95**

QUESTIONS AND ANSWERS FOR PLUMBERS EXAMINATIONS, 2nd Edition

Answers plumbers' questions about types of fixtures to use, size of pipe to install, design of systems, size and location of septic tank systems, and procedures used in installing material. 256 pages; 5½ × 8¼; softcover. **Price: $8.95**

TREE CARE MANUAL

The conscientious gardener's guide to healthy, beautiful trees. Covers planting, grafting, fertilizing, pruning, and spraying. Tells how to cope with insects, plant diseases, and environmental damage. 224 pages; 8½ × 11; softcover. **Price: $8.95**

UPHOLSTERING

Upholstering is explained for the average householder and apprentice upholsterer. From repairing and regluing of the bare frame, to the final sewing or tacking, for antiques and most modern pieces, this book covers it all. 400 pages; 5½ × 8¼; hardbound. **Price: $12.95**

WOOD FURNITURE: Finishing, Refinishing, Repair

Presents the fundamentals of furniture repair for both veneer and solid wood. Gives complete instructions on refinishing procedures, which includes stripping the old finish, sanding, selecting the finish and using wood fillers. 352 pages; 5½ × 8¼; hardbound. **Price: $9.95**

ELECTRICITY/ELECTRONICS
ELECTRICAL LIBRARY

If you are a student of electricity or a practicing electrician, here is a very important and helpful library you should consider owning. You can learn the basics of electricity, study electric motors and wiring diagrams, learn how to interpret the NEC, and prepare for the electrician's examination by using these books.

Use the order coupon on the back of this book.
All prices are subject to change without notice.

Electric Motors, 4th Edition. 528 pages; 5½ × 8¼; hardbound. **Price: $12.95**

Guide to the 1984 National Electrical Code. 672 pages; 5½ × 8¼; hardbound.
Price: $18.95

House Wiring, 6th Edition. 256 pages; 5½ × 8¼; hardbound. **Price: $12.95**

Practical Electricity, 4th Edition. 496 pages; 5½ × 8¼; hardbound. **Price: $13.95**

Questions and Answers for Electricians Examinations, 8th Edition. 288 pages; 5½ × 8¼;
hardbound. **Price: $12.95**

ELECTRICAL COURSE FOR APPRENTICES AND JOURNEYMEN, 2nd Edition
A study course for apprentice or journeymen electricians. Covers electrical theory and its applications. 448 pages; 5½ × 8¼; hardbound. **Price: $13.95**

MATHEMATICS FOR ELECTRICIANS AND ELECTRONICS TECHNICIANS
This new book offers a practical, step-by-step approach to the basic math everyone in the electri-
cal/electronics field needs to know. 312 pages; 5½ × 8¼; hardbound. **Price: $14.95**

FRACTIONAL HORSEPOWER ELECTRIC MOTORS
This new book provides guidance in the selection, installation, operation, maintenance, repair, and
replacement of the small-to-moderate size electric motors that power home appliances and over 90
percent of industrial equipment. Provides clear explanations and illustrations of both theory and
practice. 352 pages; 5½ × 8¼; hardbound. **Price: $15.95**

TELEVISION SERVICE MANUAL, 5th Edition
Provides the practical information necessary for accurate diagnosis and repair of both black-and-
white and color television receivers. 512 pages; 5½ × 8¼; hardbound. **Price: $15.95**

ENGINEERS/MECHANICS/MACHINISTS
MACHINISTS LIBRARY, 4th Edition
Covers the modern machine-shop practice. Tells how to set up and operate lathes, screw and mill-
ing machines, shapers, drill presses and all other machine tools. A complete reference library.
Price: $35.85
Volume 1
Basic Machine Shop. 352 pages; 5½ × 8¼; hardbound. **Price: $12.95**
Volume 2
Machine Shop. 480 pages; 5½ × 8¼; hardbound. **Price: $12.95**
Volume 3
Toolmakers Handy Book. 400 pages; 5½ × 8¼; hardbound. **Price: $12.95**

MECHANICAL TRADES POCKET MANUAL, 2nd Edition
Provides practical reference material for mechanical tradesmen. This handbook covers methods,
tools equipment, procedures, and much more. 256 pages; 4 × 6; softcover. **Price: $10.95**

MILLWRIGHTS AND MECHANICS GUIDE, 3rd Edition
Practical information on plant installation, operation, and maintenance for millwrights, mechanics,
maintenance men, erectors, riggers, foremen, inspectors, and superintendents. 960 pages;
5½ × 8¼; hardbound. **Price: $19.95**

POWER PLANT ENGINEERS GUIDE, 3rd Edition
The complete steam or diesel power-plant engineer's library. 816 pages; 5½ × 8¼; hardbound.
Price: $16.95

Use the order coupon on the back of this book.
All prices are subject to change without notice.

SHEET METAL WORK

A new comprehensive manual covering in illustrated detail fundamentals of sheet metal work, from basic math and drafting to layout and pattern development, materials, cutting, finishing, and much more. 456 pages; 5½ × 8¼; hardbound. **Price: $17.95**

WELDERS GUIDE, 3rd Edition

This new edition is a practical and concise manual on the theory, practical operation and maintenance of all welding machines. Fully covers both electric and oxy-gas welding. 928 pages; 5½ × 8¼; hardbound. **Price: $19.95**

WELDER/FITTERS GUIDE

Provides basic training and instruction for those wishing to become welder/fitters. Step-by-step learning sequences are presented from learning about basic tools and aids used in weldment assembly, through simple work practices, to actual fabrication of weldments. 160 pages; 8½ × 11; softcover. **Price: $7.95**

FLUID POWER
PNEUMATICS AND HYDRAULICS, 4th Edition

Fully discusses installation, operation and maintenance of both HYDRAULIC AND PNEUMATIC (air) devices. 496 pages; 5½ × 8¼; hardbound. **Price: $15.95**

PUMPS, 4th Edition

A detailed book on all types of pumps from the old-fashioned kitchen variety to the most modern types. Covers construction, application, installation, and troubleshooting. 480 pages; 5½ × 8¼; hardbound. **Price: $14.95**

HYDRAULICS FOR OFF-THE-ROAD EQUIPMENT

Everything you need to know from basic hydraulics to troubleshooting hydraulic systems on off-the-road equipment. Heavy-equipment operators, farmers, fork-lift owners and operators, mechanics—all need this practical, fully illustrated manual. 272 pages; 5½ × 8¼; hardbound. **Price: $8.95**

HOBBY
COMPLETE COURSE IN STAINED GLASS

Written by an outstanding artist in the field of stained glass, this book is dedicated to all who love the beauty of the art. Ten complete lessons describe the required materials, how to obtain them, and explicit directions for making several stained glass projects. 80 pages; 8½ × 11; softbound. **Price: $6.95**

Use the order coupon on the back of this book.
All prices are subject to change without notice.